T0270891

Theory, Techniques and Applications of Nanotechnology in Gene Silencing

Technology in Biology and Medicine

Volume 2

Series Editor

Prof. PAOLO DI NARDO

University of Rome "Tor Vergata"

Rome, Italy

In the past innovation may have been the prerogative of large, closed research labs, but their advantage over smaller rivals and the developing world is now being eroded by two powerful forces. The first is globalization, as more emerging countries become both consumers and, increasingly, suppliers of innovative products and services. The second is the rapid advance of information technologies, which are spreading far beyond the internet and into older industries such as steel, aerospace and car manufacturing. This development is being felt also in the bio-medical domain. The convergence between bio-medicine and engineering holds promises to be highly beneficial for both patients and industries. Words like information technology, advanced materials, imaging, nanotechnology and sophisticated modeling and simulation are now usual in biomedical research and clinical centers. Some scientists and analysts believe that the transformation of biology into an information science instead of a discovery science could be one of the most important innovations in the history of this science, if we are able to activate the untapped ingenuity of people and to accept the risk and the possible failure as central factor of innovation.

The info-medical revolution will be led by information technologies through the establishment of an intelligent network that will enable many other big technological changes. The progressive increase of degenerative diseases and related disabilities together with the extraordinary expansion of longevity will determine the impossibility to sustain health system overspending in most countries. Info-medicine can enhance the quality and quantity of health care, creating equal access to the most advanced diagnostic and treatment procedures by all individuals independently of geographic and economic conditions. The adoption of these technologies, however, depends on absolute proof that they will produce better outcomes and offer value for money.

It is TBM's ambition to represent an international forum to discuss in-depth the most advanced concepts and solutions merging biology, medicine and engineering. To this end, the journal will give scientists the opportunity to publish their most important results reaching the widest audience, while the book series will thoroughly dissect the new concepts in order to supply students and professionals with the most advanced tools to apply them, all in the shortest possible time frame.

For a list of other books in this series, please visit www.riverpublishers.com
For a list of other books in this series, see final page.

Theory, Techniques and Applications of Nanotechnology in Gene Silencing

Dr. Surendra Nimesh

Prof. Ramesh Chandra

River Publishers

Aalborg

Published, sold and distributed by:
River Publishers
PO box 1657
Algade 42
9000 Aalborg
Denmark
Tel.: +4536953197

www.riverpublishers.com

ISBN: 978-87-92329-83-7
© 2011 River Publishers

Foreword

"Theory, Techniques and Applications of Nanotechnology in Gene Silencing" is a timely book prepared by Surendra Nimesh and Ramesh Chandra. It describes one of the many important uses of nanotechnology. The authors first describe the basic science related to siRNA technology and strategies to overcome problems related to *in vitro* and *in vivo* siRNA delivery. This is followed by an excellent overview of the available published data for both *in vitro* and *in vivo* gene silencing mediated by polymeric nanoparticles and polymers used for the preparation of polymeric nanoparticles to deliver siRNA to mediate gene silencing. This book also provides an indepth overview of the more specialized area related to Nanotechnology in Gene Silencing. Nanotechnology is now a large area and my foreword would add some more general points related to nanotechnology and nanobiotechnology.

Nanotechnology includes membranes of nanodimension thickness, particles of nanodimension diameter or nanodimension molecular assembly. A subdivision of nanotechnology is nanobiotechnology which is the assembling of biological molecules into membranes of nanodimension thickness or particles of nanodimension diameters. Artificial cells with nanodimension thickness polymeric membranes were prepared as early as 1957 [1, 2]. This included the use of nanobiotechnology to prepare artificial cells with nanodimension thickness crosslinked with protein membrane or protein-polymer conjugated membranes [2]. These nanobiotechnological membranes are biodegradable. Artificial cells with nanothickness polymeric membranes prepared from polylactic acid [3] or copolymer of polyethylene glycol-polylactic acid [4]are also biodegradable. Artificial cells can have unlimited compositions and contents. Examples of such contents include the magnetic material, enzymes, hemoglobin, genes, drugs, hormone, antibody, antigenetc [5–7]. The original artificial cells are of micron dimensions of down to one micron diameter.

However, by slight modification of the original emulsion method, nanodimension artificial cells or nanoparticles and other nanodimension composition can be easily prepared [6, 7]. These are now called by many different names: Nanocapsules, nanoparticles, liposomes, polymersomes,nanorods, nanotubes, cones, spheroids, etc. [7]. In addition, biological molecules like enzymes and hemoglobin can be assembled into soluble nanodimension structures for use as blood substitutes or oxygen carriers with therapeutic properties [8, 9].

The above is only a summary of some aspects of nanotechnology in nanomedicine. More details are available in Reference [10].

References

[1] Chang TMS. Hemoglobin corpuscles. *Honours Physiology Research Report*, Medical Library, McGill University, 1957. Also reprinted as part of 30th anniversary in artificial red blood cells research. *BiomaterArtif Cells Artif Organs*, 16: 1–9, 1988.

[2] Chang TMS. Semipermeable microcapsules. *Science*, 46(3643): 524–515, 1964.

[3] Chang TMS. Biodegradable semipermeable microcapsules containing enzymes, hormones,vaccines, and other biologicals. *Journal of Bioengineering*, 25–32, 1976.

[4] Chang TMS., Powanda D. and Yu WP. Ultrathin polyethylene-glycol-polylactide copolymer membrane nanocapsules containing polymerized Hb and enzymes as nano-dimension RBC substitutes. *Artificial Cells, Blood Substitutes and Biotechnology*, 31: 231–248, 2003.

[5] Chang TMS. *Artificial Cells*. C Thomas, Springfield, IL, 1972(full text available at www.artcell.mcgill.ca).

[6] Chang TMS. Therapeutic applications of polymeric artificial cells. *Nature Review: Drug Discovery*, 4: 221–235, 2005 (full text on www.artcell.mcgill.ca).

[7] Chang TMS. *ARTIFICIAL CELLS*: *Biotechnology, Nanotechnology, Blood Substitutes, Regenerative Medicine, Bioencapsulation, Cell/Stem Cell Therapy.* World Science Publisher, Singapore, pp. 1–452, 2007 (full text on www.artcell.mcgill.ca).

[8] D'Agnillo, F. and TMS Chang. Polyhemoglobin-superoxide dismutase-catalase as a blood substitute with antioxidant properties. *Nature Biotechnology*, 16(7): 667–671, 1998.

[9] Chang TMS. Blood replacement with engineered hemoglobin and hemoglobin nanocapsules, *Interdisciplinary Reviews on Nanomedicine and Nanobiotechnology*, 2: 418–430, 2010.

[10] Website of Artificial Cells and Organs Research Centre of McGill University on *Artificial Cells, Blood Substitutes and Nanomedicine.* www.artcell.mcgill.ca

Professor Thomas Ming Swi Chang, OC,MD,CM,PhD,FRCPC.FRS[C]

Director, Artificial Cells and Organs Research Centre

Departments of Physiology, Medicine and Biomedical Engineering

Faculty of Medicine, McGill University

Montreal, Quebec, Canada H3G1Y6

Artcell.med@mcgill.ca

Preface

Nobel Laureate Richard P. Feynman gave a talk at the annual meeting of the American Physical Society on Dec. 29, 1959, at the California Institute of Technology, that has become one of the twentieth century's classic science lectures, titled "There's Plenty of Room at the Bottom". He emphasized on a technological vision to manipulate and control things on miniature scale. Feynman envisaged a technology of building nano-objects atom by atom or molecule by molecule. Since the 1980s, several inventions and discoveries in the fabrication of nano-objects have realized his vision. The term "nanotechnology" was coined by Tokyo Science University Prof. Norio Taniguchi in 1974 and defined as: Nano-technology mainly consists of the processing of separation, consolidation, and deformation of materials by one atom or one molecule. Nanotechnology enables the production and application of physical, chemical, and biological systems at sizes ranging from individual atoms or molecules to submicron dimensions. Innovation in nanotechnology promises breakthroughs in areas such as materials and manufacturing, nanoelectronics, medicine and healthcare, energy, biotechnology, information technology, and national security. A unique parameter of nanotechnology is the highly increased ratio of surface area to volume present in many nanoscale materials which opens new avenues in surface-based science, such as catalysis. A number of physical properties become noticeably visible as the size of the system decreases. These aspects include statistical mechanical effects, as well as quantum mechanical effects, for example the "quantum size effect" where the electronic properties of solids are altered with great reductions in particle size.

This book on "Theory, Techniques and Applications of Nanotechnology in Gene Silencing" highlights the recent past advances at the interface between the science and technology of nanotechnology and their biological applications. As highlighted in this book these applications include using nanoparticles for *in vitro* and *in vivo* siRNA delivery for gene silencing. The potential implications of nanotechnology have been demonstrated recently by a broad variety of applications in the study of sub-cellular processes of fundamental importance in biological systems. Examples include but not limited to study of mechanism, diagnosis, treatment of deadly diseases such as cancer. Chapter 1 provides an introduction to nanotechnology detailing the evolution of nanoparticles as distinguished drug and gene delivery systems. It also described the potential importance and application of nanotechnology in drug, gene and siRNA delivery. Chapters 2, 3 details about the various tools and techniques employed to study nanoparticles physicochemical properties and the role therein in investigation for delivery applications. It provides an overall aspect of various strategies adapted for tracing the movement of nanoparticles while in various *in vitro* and *in vivo* applications.

The success of nanoparticles mediated siRNA delivery and henceforth gene silencing is hampered by various *in vitro* and *in vivo* obstacles. The mechanism of RNAi takes place in the cytoplasm of the cells and is a multi-step process; Chapter 4 describes the mechanism of siRNA mediated gene silencing i.e., RNA interference. It also gives an account of the barriers encountered and the strategies adapted to successfully combat them. The barriers have been sub divided into two classes i.e., the *in vitro* and *in vivo* barriers. Although modifications such as PEGylation have improved the circulation time of nanoparticles, the RES still manages to capture a significant amount of the injected dose. Henceforth, a subtle balance between stability and cellular uptake is desirable for siRNA to meet its potential as a therapeutic modality. The potential of nanoparticles can further be manipulated by making them target specific by decoration of surface with various targeting ligands. Chapter 5 mentions the cell surface modifications that occurs in the diseased cells and the strategies designed to exploit that situation by employing different targeting ligands to nanoparticles such as antibodies, transferrin etc. Nanoparticles, being compact, are well suited to traverse cellular membranes to mediate gene delivery. It is also expected that due to smaller size, nanoparticles would be less susceptible to reticuloendothelial system clearance and will have better penetration into tissues and cells, when used in *in vivo* therapy. Chapter 6 highlights the pros and cons of using cationic polymeric nanoparticles for siRNA delivery.

Chitosan, an aminoglucopyran, which is biocompatible and biodegradable along with low toxicity, has been largely explored as siRNA delivery candidate. It has been employed in numerous *in vitro* and *in vivo* studies for gene silencing applications in cell and animal models. Chapter 7 provides account of properties of chitosan and preparation methods in addition to preparation of nanoparticles for siRNA delivery. Polyethylenimine (PEI) often considered as the gold standard of gene transfection have recently been investigated as a siRNA delivery vector. PEI exists in different molecular weights and polymer isoforms. Chapter 8 provides brief summary of the studies undertaken to investigate the siRNA delivery potential of PEI. Poly-L-lysine (PLL) is one of the first polymers investigated for non-viral gene delivery, but found limited applications for siRNA. Chapter 9 explores the literature citing implications of PLL for siRNA delivery. Atelocollagen was the first biomaterial with potential application as a gene delivery vector and obtained by pepsin treatment from type I collagen of calf dermis. siRNA complexed with atelocollagen is resistant to nucleases and is transduced efficiently into cells, thereby allowing long-term gene silencing. Chapter 10 details the studies done to tap the hidden potential of atelocollagen for gene delivery in several mice models.

Use of protamine for siRNA delivery was first reported by Song *et al.* which further followed by several studies. Chapter 11 deals with application of protamine in different formulations to provide safe and effective siRNA delivery. Dendrimers, the branched, multivalent molecules with a well defined structure found ample of applications in gene delivery and also explored for siRNA delivery. Chapter 12 summarizes different types of dendrimers, structural arrangements and their siRNA delivery potential. Cyclodextrin containing cationic polymer (CDP) for gene delivery can be further modified by inclusion-complex-formation due to presence of large amounts of CD moieties. Chapter 13 reports the use cationic polymers such as linear and branched PEI, and PAMAM dentrimer, modified by grafting CDs on the polymers, and studied for siRNA delivery. Although, cationic polymers (i.e., nucleic acids condensing agents) seem to be an excellent substitute for viral vectors, limitations including toxicity due to excess positive charge of the polymer, limits the application for systemic delivery. Chapter 14 explores the promising

PLGA-based nanoparticles with the encapsulation of siRNA, for delivery to plethora of applications. Nanomaterials such as gold nanoparticles, carbon nanotubes, and silica nanoparticles attracted considerable attention in their pharmaceutical applications due to their biocompatibility and ability to facilitate the delivery of therapeutic cargos. Chapter 15 discusses the properties and use of these nanomaterials for siRNA delivery.

Collectively, this book should prove a useful resource not only to those who are using or wish to use nanoparticles for siRNA delivery in cell lines and animal models, but also to those interested in applications of nanotechnology more generally. We hope that this book will also serve as a catalyst spurring others to use nanomaterials not yet being widely studied for siRNA delivery and to develop new methodologies that would contribute to both the fundamental understanding of nanomaterials mechanism of action and their potential utility.

Acknowledgements

A work of this scope would be impossible without the dedication and hard work of many people. We sincerely thank the scientists who have devoted their career for understanding the nanotechnology world, and provide careful explanation and clarity enough for a beginning audience. We wish to express our gratitude to those who generously helped to color the mosaic of this book on "Theory, Techniques and Applications of Nanotechnology in Gene Silencing" with the tiles of their knowledge and expertise.

The first and foremost thanks are due to Prof. T. M. S. Chang, FRS, Director, Artificial Cells and Organs Research Centre, McGill University, Montreal, Canada, for his persistent interests and generous availability of all his expertise in the field of nanotechnology throughout the course of this work. His optimism and dynamic attitude have been a constant source of inspiration.

The next best but equally heartfelt gratitude goes to Prof. Satya Prakash, Biomedical Engineering Technology and Cell Therapy Research Laboratory, McGill University, Montreal, Canada, for his encouragement, motivation, undying patience and critical evaluation are too tremendous to fit into words. He has been a guiding star all along, supporting with his inexhaustible knowledge and immense experience. His invaluable inputs and unfailing encouragement throughout this work is an asset.

We are deeply indebted to Prof. Sukh Dev, FNA, INSA-S. N. Bose Research Professor, and Dr. B. R. Ambedkar Center for Biomedical Research, University of Delhi, Delhi, India for his expert guidance and comments during the course of this work.

Special thanks are due to Prof. Goverdhan Mehta, FRS, National Research Professor, and Lilly-Jubilant Chair Professor at University of Hyderabad, Hyderabad, India, also a member of the Scientific Advisory Committee to the Prime Minister of India, and Former President of the Indian National Science Academy and International Council for Science (ICSU), for critical examination of the manuscript, sage advice, useful comments and encouragement, which have greatly enhanced the quality, presentation and content of this work.

Carefully reviewing a book means lots of work but not much appreciation for the reviewers. Therefore, the authors wish to express their deep gratitude to Dr. Nidhi Gupta and Dr. Vineet Agrawal, Reddy's Laboratories, Hyderabad, India, for taking on this difficult job.

The authors shall never be able to express, to any degree of satisfaction, their deepest feelings of gratefulness to their friends and whole family for their untiring efforts and for helping to finalize this work. We also wish to express our appreciation to our many associates and colleagues, who, as experts in their fields, have helped us with constructive criticism and helpful suggestions.

Finally, we are indebted and would like to thank the outstanding members of the management of River Publishers, Denmark for supporting the publication of this book and have been of enormous help in bringing this edition to fruition in a timely manner.

Dr. Surendra Nimesh, Ph.D.
Prof. Ramesh Chandra, Ph.D, FRSC

Contents

Abbreviations

3-(4,5-dimethylthiazol-2-yl)-2,5-diphenyl tetrazolium bromide	MTT
Acrylamido-2-deoxy-glucose	AADG
Antibody(s)	Ab(s)
Apolipoprotein	Apo
Arg-Gly-Asp	RGD
Asialoglycoprotein receptor	ASGP-R
Atomic force microscope	AFM
Bovine serum albumin	BSA
Branched Polyethylenimine	BPEI
Carbon nanotubes	CNTs
Cell penetrating peptides	CPPs
Cyclodextrin	CD
Cyclodextrin containing cationic polymer	CDP
Degree of deacetylation	DDA
Degree of polymerization	DP
Double-stranded RNA	dsRNA
Dynamic light scattering	DLS
Enhanced green fluorescent protein	EGFP
Epidermal growth factor receptor	EGFR
Fluorescence assisted cell sorting	FACS
Folate receptors	FRs
Gold nanoparticles	AuNPs
Green fluorescent protein	GFP
Human serum albumin	HSA
Isothermal titration calorimetry	ITC

Linear polyethylenimine	LPEI
Liposome-polycation-hyaluronic acid	LPH
M2 subunit of ribonucleotide reductase	RRM2
Mesoporous silica nanoparticles	MSNs
Mitogen-activated protein kinase 1	MAPK1
Molecular weight	MW
Monoclonal antibodies	mAbs
Mononuclear phagocyte system	MPS
Multi-Angle Laser Light Scattering	MALLS
Multidrug resistance	MDR
Multiple walled carbon nanotubes	MWCNTs
N,N-methylene bis-acrylamide	MBA
N-isopropylacrylamide	NIPAM
Nitrogen of amine to phosphate of DNA/siRNA	N/P ratio
N-tert-butylacrylamide	BAM
Phosphate buffered saline	PBS
Photochemical internalization	PCI
Platelet-derived growth factor BB	PDGF-BB
Poly(amidoamine)	PAMAM
Poly(propylenimine)	PPI
Poly(vinyl alcohol)	PVA
Poly(β-amino ester)s	PBAEs
Polyacrylamide gel electrophoresis	PAGE
poly-D,L-lactide-co-glycolide	PLGA
polydispersity index	PDI
Polyethylene glycol	PEG
Polyethylene oxide	PEO
Polyethylenimine	PEI
Polyhexadecylcyanoacrylate	PHDCA
Polyion complex	PIC
Poly-L-lactide	PLA
Poly-L-lysine	PLL
Polymerase chain reaction	PCR
Quantitative reverese transcription polymerase chain reaction	qRT-PCR
Red fluorescent protein	RFP
Reticuloendothelial system	RES

Retinal pigment epithelium	RPE
RNA interference	RNAi
Scanning electron microscope	SEM
Scanning force microscope	SFM
Single chain variable antibody fragment	scFV
Single walled carbon nanotubes	SWCNTs
Small interfering RNA	siRNA
Sodium dodecyl sulfate	SDS
Thiol	SH
Transferrin receptor	TFR
Transmission electron microscopy	TEM
Tripolyphosphate	TPP
Vascular endothelial growth factor	VEGF
Vascular endothelial growth factor receptor-2	VEGFR2

1

Introduction to Nanotechnology

1.1 Introduction

The concept of a "magic bullet" drug was first emphasized by the German scientist Paul Ehrlich in the 19th century, while studying selective staining of tissues for histological examination, in particular, selective staining of bacteria. Ehrlich envisaged that if a compound could be developed to selectively target the disease causing organism, then a toxin for that organism could be delivered along with the agent of selectivity. Hence, a "magic bullet" could be designed that only targets and kills the disease causing organism. The concept offered numerous advantages over the traditional delivery of drugs to specific sites, i.e.

  (i) the delivery of drug was site specific i.e. drug was delivered at the diseased site,
 (ii) several inaccessible diseased sites were now made accessible,
(iii) undesired effects of the delivered drug could be minimized, and
(iv) controlled delivery of drugs i.e. lower amounts of drugs was sufficient to bring the desired therapeutic effect.

During the past few decades, a number of strategies have been proposed to develop an efficient drug delivery system for required therapeutic effect. The physicochemical properties and molecular structure of the drug determine the fate of its distribution in the body upon administration by various available routes. Usually, due to non-optimal distribution of the drug, only a small amount of the administered dose reaches the desired diseased site for action.

S. Nimesh and R. Chandra,
Theory, Techniques and Applications of Nanotechnology in Gene Silencing, 1–9.
© 2011 *River Publishers. All rights reserved.*

Accumulation of the drugs at nonspecific sites may lead to adverse and toxic side effects. Thus, it has been a challenging task for the scientists to develop delivery systems having maximized drug action and diminished side effects. To achieve accurate drug targeting, the description regarding the nature of the target and the mechanism of targeting should be well understood. A number of drug delivery systems have been developed that can efficiently and specifically deliver drugs to target sites. The prodrugs i.e. the modified forms of the active drugs that can reach the target sites and be cleaved enzymatically or chemically to release the active drug moiety are one of the chemically evolved approaches. A number of bioconjugate-based systems have been studied, including the use of monoclonal antibodies (mAbs), polyclonal antibodies, sugars or lectins as targeting moieties, where a conjugate of the drug can be formed by coupling it chemically with these moieties.

The choice of a drug delivery system plays an important role in achieving site specific delivery of a drug. In the recent past, a number of carrier systems have been proposed such as: liposomes, nanoparticles, microparticles, and microemulsions. These systems have been developed to prevent any interactions with non-target sites, resulting in the localization of the drug at the target sites. Carrier conjugated drug delivery system offers distinct advantages such as:

    (i) active form of the drug can be maintained,
   (ii) therapeutic concentration can be maintained at diseased sites,
  (iii) minimal exposure of drug to normal metabolic and immune attack; subsequent decrease in renal excretion leads to increased half-life of drug,
  (iv) activation of the drug at target site can be achieved,
   (v) cell specific interaction can be achieved by overcoming numerous physiological barriers.

The foremost goal of a controlled drug delivery is to deliver a particular drug to the target site, or a specific population of cells within a particular organ or tissue. For designing a new system for the target specific delivery of a drug, the following factors should be taken into consideration:

    (i) in order to minimize the toxic effects, the system should be able to differentiate between target and non-target sites,

(ii) no toxic effect should be observed on prolonged retention in the system,

(iii) the carrier system should be biodegradable i.e. it should be decomposed/degraded after delivering the drug to the target site and the subsequent degraded components should not be toxic to the system in which it is administered.

The systemic scavenging machinery such as the reticuloendothelial system (RES) has posed as one of the major challenges in the development of polymeric nanoparticulate carriers. When the carrier molecules or any other foreign moiety enter the vasculature system, they are rapidly conditioned (or coated) by elements of circulation, such as plasma proteins and glycoproteins called as "opsonin", by a process known as "opsonization". In this opsonization process, the carrier materials are easily recognized by RES and like any other foreign bodies, pathogens or dying cells, they are recognized as foreign products and removed from the blood circulation by phagocytosis [1]. The macrophage cells of the liver (Kupffer cells), spleen, lung, and circulating macrophages, all play a vital role in removing the opsonized particles. The size and the surface properties of the particle are significant for the recognition of the particle by RES. Particulates with large hydrophobic surfaces are efficiently coated with opsonins and are rapidly cleared from circulation. However, particles with hydrophilic surfaces escape being recognized by opsonins which leads to prolonged circulation [2].

Wilkins *et al.* developed different strategies for altering the surface properties of carrier molecules [3]. A reduction in the liver uptake and an increase in the spleen and lung uptake were observed on coating polystyrene nanoparticles with positively charged poly-L-lysine (PLL) gelatin. In the early 1980's Illum *et al.* discovered that it was possible for the particles to escape from RES scavenging to a great extent when the surface characteristics of the particles was changed by coating them with block copolymers [2]. This strategy has been shown to be highly effective in altering the biodistribution pattern of radiolabeled colloidal particles. The polystyrene particles coated with block copolymer poloxamer 338, greatly escaped (upto 50%) the normally predominant liver and spleen uptake; however, these particles were rapidly cleared by RES cells. The clearance could be prevented by coating the particles with poloxamine 908, which totally prevents the RES capture, leading to prolonged circulation times. Some of the other strategies proposed to circumvent these problems include coating the carrier molecules with surfactants, gangliosides and polymers such as polyethylene glycol (PEG) or polyethylene oxide (PEO). Studies with surfactant coated carrier molecules led to reduced liver uptake and increase in their concentration in blood and other non-RES organs [4]. However, such strategies are not considered to be applicable in clinical practice, since repeated suppression can lead to an impaired RES function.

During the last few years, genes have become potential targets for therapeutic use in a wide variety of disease states due to their ability to produce active proteins using the biosynthetic machinery provided by the host cells. The basic principle of "gene therapy" lies in the insertion of genetic material, i.e., DNA or RNA, into the cell's genetic machinery either to correct an underlying defect or to modify the characteristics of the cell via expression of the newly inserted gene. For efficient delivery of genes to the target cells, various vectors have been developed, which can be categorized into viral and nonviral vectors. The viral vectors are potential carriers due to site specificity and the natural mechanism employed to enter the cell, but their large scale use is hampered due to immunogenicity and pathogenicity [5]. A large number of nonviral vectors including lipid-based carriers, polycationic lipids, polylysine, polyornithine, histones and other chromosomal proteins, hydrogel polymers, precipitated calcium phosphate and calcium phosphate nanoparticles have been developed. One of the major limiting factors behind the use of nonviral vectors particularly polymeric nanoparticles is their extraordinary stability in the endosomal compartment, which can lead to the destruction of DNA by the lysosomal enzymes. Highly stable complexes not only make DNA more prone to enzymatic degradation but will not allow the release of DNA from the endosome into the cytosol, resulting in low transfection efficiency.

For enhancing the bioavailability of a drug at the desired site and for exploiting the therapeutic aims of the controlled release of a drug and drug targeting, the particle-based drug carriers have been developed as an improved and alternative method. Controlling the size of polymeric matrix is highly desirable to achieve the desired therapeutic response of the entrapped biomolecules. On the basis of size, the particles can be grouped into three broad categories:

(i) macroparticles (50–200 μm)
(ii) microparticles (1–50 μm)
(iii) nanoparticles (1–1000 nm)

Among these, microparticles and nanoparticles have been used for therapeutic purposes. Microparticles can circulate in the blood and also pass through the heart but cannot enter capillaries and thus cannot reach tissue sinusoids, because of their large size. Drug encapsulated molecules get accumulated in the nearby tissues surrounding the capillaries and release the drug slowly. The delivery of these microparticles can be achieved by intraarterial, subcutaneous, intravenous or intraperitoneal routes and these accumulated microparticles can be used as depot system. In this way, a drug is released slowly but continuously and simultaneously and in the process gets protection from degradation. The drug release is monitored by selecting a properly designed polymer, its pore size, swelling properties and degradation. Keeping these properties in mind, researchers have prepared biodegradable microparticles of starch, albumin, polylactic acid and ethyl cellulose and used them for chemoembolisation.

Further drug delivery studies have revealed that the micron sized particles are rapidly cleared by RES cells. To prevent this rapid removal of particles from the circulation, a decrease in the size of the particles was thought to help in prolonged circulation and would constitute a better carrier for enzymes, proteins, polynucleotides by any route of administration. Several attempts were made to prepare submicron sized particles to develop carrier systems to deliver a specific drug molecule at a pre-determined target site. Furthermore, these particles would provide more surface area for larger amounts of the drug or biological molecule to be adsorbed. Subsequently, it has been demonstrated that nanoparticles have the capability of not only releasing the drug or a bioactive molecule at the target site but also carrying it there. The size of the particle made huge difference and rendered them suitable for systematic use, as they can pass through capillaries. Nanoparticles, as the name denotes, are the metallic or polymeric particles with the size range between 1–1000 nm. The first report on the preparation and characterization of the polymeric nanoparticles was published in 1976 by Birrenbach *et al.* [6]. Since then, this area of research has grown up explosively and started showing up significant commercial impact. Polymeric nanoparticles carrying drug/biomolecule deliver the same at the target site via one of the following mechanisms:

(i) by swelling the polymeric nanoparticles with hydration followed by their release through diffusion,

(ii) by an enzymatic reaction leading to the degradation of the polymeric network at the target site, thereby releasing the drug, and

(iii) by cleavage of the drug from the swelled nanoparticles.

1.2 Definition of Nanotechnology

Nobel Laureate Richard P. Feynman gave a talk at the annual meeting of the American Physical Society on December 29, 1959, at the California Institute of Technology, that has become one of the twentieth century's classic science lectures, titled "There's Plenty of Room at the Bottom". He emphasized on a technological vision to manipulate and control things on miniature scale. Feynman envisaged a technology for building nano-objects atom-by-atom or molecule- by-molecule. Since the 1980s, several inventions and discoveries in the fabrication of nano-objects have realized his vision. The term "nanotechnology" was coined by Tokyo Science University Prof. Norio Taniguchi in 1974 and defined thus: Nanotechnology mainly consists of the processing of separation, consolidation, and deformation of materials by one atom or one molecule. In addition, nanotechnology refers to technology performed on a nanoscale that has potential applications (Figure 1.1). Nanotechnology enables the production and application of physical, chemical, and biological systems at sizes ranging from individual atoms or molecules to sub-micron dimensions. Innovation in nanotechnology promises breakthrough in areas such as materials and manufacturing, nanoelectronics, medicine and healthcare, energy, biotechnology, information technology, and national security. A unique parameter of nanotechnology is the highly increased ratio of surface area to volume present in many nanoscale materials which opens new avenues in surface-based science, such as catalysis. A number of physical properties become noticeably visible as the size of the system decreases. These aspects include statistical mechanical effects, as well as quantum mechanical effects,

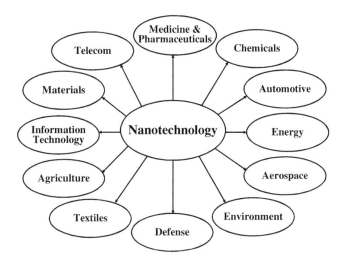

Fig. 1.1 Applications of nanotechnology.

for example the "quantum size effect" where the electronic properties of solids are altered with great reductions in particle size.

Nanomedicine is one of the most important applications of nanotechnology; it is used to collectively mention polymeric micelles, quantum dots, liposomes, polymer-drug conjugates, dendrimers, biodegradable nanoparticles, inorganic nanoparticles and other materials in nanometer size range with therapeutic relevance. Nanoparticles have emerged as one of the most promising candidates of nanomedicine, with numerous applications in targeted drug and gene delivery. Owing to their small size, nanoparticles provide better tissue penetration and targeting [7]. Nanoparticles prepared from polycationic polymers have been largely explored to deliver DNA and Small Interfering RNA (siRNA). These nanoparticles can be subdivided as:

(i) Nanospheres: These are spherical nanometer size particles where the desired molecules can be either entrapped inside the sphere or adsorbed on the outer surface or both.
(ii) Nanocapsules: These have a solid polymeric shell and an inner liquid core where the desired molecules can be entrapped (Figure 1.2).

However, nanoparticles have also been reported to exist in different types of shapes such as nanorods, nanotubes, cones, spheroids etc. Recently,

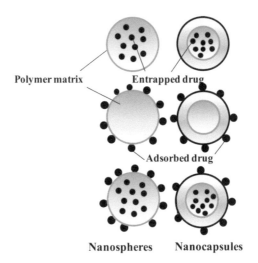

Fig. 1.2　Different types of nanoparticles.

carbon nanotubes have also been conjugated with polyglycerol to prepare nanocapsules [8]. Several natural and synthetic polycationic polymers including chitosan, polyethylenimine (PEI) have been utilized for preparing nanoparticles to deliver siRNA. Herein, we have focused on recent strategies and advances in the application of polymeric nanoparticles for targeted siRNA delivery.

1.3　Composition of the Book

The book integrates knowledge of basic concepts in preparation, characterization and biological applications of polymeric nanoparticles. The first six chapters provide indepth knowledge about the various tools available for identification of nanoparticles followed by basic science involved behind the siRNA technology. These chapters provide details about the major barriers encountered during *in vitro* and *in vivo* siRNA delivery and the strategies adapted to overcome them. The second half of the book provides an overview of the various polymers such as chitosan, PEI, polylysine, cyclodextrin, etc. employed to deliver siRNA to mediate gene silencing. An emphasis has been given on the incorporation of most up-to-date available published data for both *in vitro* and *in vivo* gene silencing mediated by polymeric nanoparticles.

References

[1] Müller, R.H. and K.H. Wallis. Surface modification of i.v. injectable biodegradable nanoparticles with poloxamer polymers and poloxamine 908. *International Journal of Pharmaceutics*, 89: 25–31, 1993.

[2] Illum, L. and S.S. Davis. The organ uptake of intravenously administered colloidal particles can be altered using a nonionic surfactant (Poloxamer 338). *FEBS Letters*, 167: 79–82, 1984.

[3] Wilkins, D.J. and P.A. Myers. Studies on the relationship between the electrophoretic properties of colloids and their blood clearance and organ distribution in the rat. *British Journal of Experimental Pathology*, 47: 568–576, 1966.

[4] Vittaz, M., D. Bazile, G. Spenlehauer, T. Verrecchia, M. Veillard, F. Puisieux, and D. Labarre. Effect of PEO surface density on long-circulating PLA-PEO nanoparticles which are very low complement activators. *Biomaterials*, 17: 1575–1581, 1996.

[5] Yang, Y., F.A. Nunes, K. Berencsi, E.E. Furth, E. Gonczol, and J.M. Wilson. Cellular immunity to viral antigens limits E1-deleted adenoviruses for gene therapy. *Proceedings of the National Academy of Sciences of the United States of America*, 91: 4407–4411, 1994.

[6] Birrenbach, G. and P.P. Speiser. Polymerized micelles and their use as adjuvants in immunology. *Journal of Pharmaceutical Sciences*, 65: 1763–1766, 1976.

[7] Peer, D., J.M. Karp, S. Hong, O.C. Farokhzad, R. Margalit, and R. Langer. Nanocarriers as an emerging platform for cancer therapy. *Nature Nanotechnology*, 2: 751–760, 2007.

[8] Adeli, M., N. Mirab, and F. Zabihi. Nanocapsules based on carbon nanotubes-graft-polyglycerol hybrid materials. *Nanotechnology*, 20: 485603, 2009.

2

Physicochemical Characterization
of Nanoparticles

2.1 Introduction

The physicochemical properties of nanoparticles, namely size, shape and surface chemistry, are well correlated with their therapeutic efficacy. A plethora
of studies have correlated the size and shape of nanoparticles with their drug
and gene delivery properties [1, 2]. Till date, numerous techniques have been
made available to quantify these properties, several of which are summarized
in this chapter. Although, the physicochemical characterization of nanoparticles are necessary for a comprehensive analysis of its biological interactions
and uptake by cells, such measurements are insufficient in predicting toxicity
of these nanoparticles.

2.2 Physicochemical Characterization

Nanoparticles possessing pharmaceutical importance were initially prepared by Birrenbach *et al.* in 1976 [3]. They synthesized polyacry-
lamide nanoparticles by the polymerization of acrylamide crosslinked
with N, N'-methylenebisacrylamide (MBA) in an inverse microemulsion
(water-in-oil) reaction. Kreuter *et al.* used another methodology called dispersion polymerization for the preparation of poly (methyl methacrylate)
nanoparticles [4]. Later, a number of methodologies were developed for the
preparation of nanoparticles entrapping desired biomolecules. The nanoparticles thus prepared have found a large number of applications in the field of
drug and gene delivery. Polymeric nanoparticles can be synthesized in two

S. Nimesh and R. Chandra,
Theory, Techniques and Applications of Nanotechnology in Gene Silencing, 11–25.
© 2011 *River Publishers. All rights reserved.*

different ways, i.e. from the preformed polymers or by polymerization of monomers [5]. Till date a large number of methodologies have been proposed for the preparation of nanoparticles depending upon the type of molecule to be delivered. Various natural and/or synthetic polymers have been employed in the preparation of polymeric nanoparticles-based therapeutics, including PEG, N-(2-hydroxypropyl) methacrylamide copolymers, poly(vinylpyrrolidone), PEI, chitosan, hyaluronic acid, dextran and poly(aspartic acid) [6]. The various techniques employed for the characterization of nanoparticles have been discussed in the following sections.

2.2.1   Size and Size Distribution

The size of nanoparticles is one of the critical factors that could affect its transfectivity. The particle size is emphasized while preparing various formulations systems such as DNA/polymer, lipid complexes and liposomes [7–9]. Several studies have shown that the particle size significantly affects their cellular and tissue uptake, and in some cell lines only the sub-micron size particles are taken up efficiently but not the larger size microparticles (e.g. Hepa 1–6, HepG2, and KLN 205) [10, 11]. Prabha *et al.* reported that the smaller-sized nanoparticles (mean diameter = 70 ± 2 nm) showed a 27-fold higher transfection than the larger-sized nanoparticles (mean diameter = 202 ± 9 nm) in COS-7 cell line and a 4-fold higher transfection in HEK-293 cell line [12]. In a study by Chithrani *et al.* the uptake of 14-, 50-, and 74-nm gold nanoparticles was investigated in HeLa cells [13, 14]. It was found that the kinetics of uptake as well as the saturation concentration varied with the different sized nanoparticles with 50 nm size particles being the most efficient in their uptake, indicating that there might be an optimal size for efficient nanomaterial uptake into cells. The effect of nanoparticle shape on its internalization was also examined: Spherical particles of similar size were taken up 500% more than rod-shaped particles, which is explained by greater membrane wrapping time required for the elongated particles. Studies involving nanoparticle size determination are often complicated by the polydispersity of samples. However, characterization of nanoparticles-employing multiple techniques, such as transmission electron microscopy (TEM) and dynamic light scattering (DLS) provides size information on nanoparticles which may be relevant to their therapeutic potential.

2.2.1.1   Microscopy

Since nanoparticles are very small in size, they are usually irresolvable by optical microscopy; hence better resolution techniques such as electron microscopy are required for determining their size and shape. TEM employs a beam of electrons which is transmitted through an ultra thin specimen, interacting with the specimen as it passes through. This interaction of electrons forms an image which is further magnified and can be focused onto an imaging device, such as a fluorescent screen, or CCD camera. Samples for nanoparticle's characterization can be either prepared by simple deposition of dilute suspensions on copper grids or can be fixated using a negative staining material such as uranyl acetate. TEM not only provides the size of nanoparticles but also furnishes information about the surface morphology. In one of our studies, samples for TEM were prepared by dispersing lyophilized powder (2 mg) of nanoparticles by sonication in distilled water (1 ml) to obtain a clear suspension [15]. The sample solution ($3 \mu l$) was put on a formvar (polyvinyl formal) coated copper grid and air-dried prior to analysis. TEM analysis of crosslinked acrylamido-2-deoxy-glucose (AADG) nanoparticles' images showed spherical and compact particles with average size of 85 nm (Figure 2.1). In another study, prior to visualization of samples, the grids were negatively stained with saturated solution of uranyl acetate [16].

Atomic force microscope (AFM) or scanning force microscope (SFM) is a very high resolution type of scanning probe microscope with resolution in the range of nanometers. The AFM consists of a cantilever with a very fine tip (probe) at its end, typically silicon with radius of few nanometers, which is used to scan the specimen surface. While scanning, the forces between the tip and the sample lead to deflection of the cantilever which can be measured using a laser spot reflected from the top surface of cantilever. The AFM can be employed to image and manipulate atoms and structures on a variety of surfaces. AFM is ideally suited for characterizing nanoparticles as it offers the capability of 3D visualization and both qualitative and quantitative information of the sample topology including morphology, surface texture and roughness and more importantly, in this case, the size of the particles [17]. In recent times, AFM has been used to determine the size and surface morphology of nanoparticles. In one of our studies, lyophilized powder ($\sim$0.5 mg) of nanoparticles was dispersed by sonication in double distilled water (1 ml)

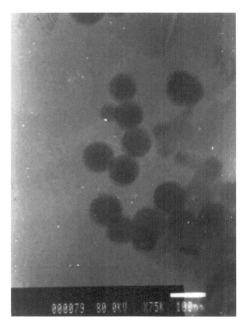

Fig. 2.1 TEM image of AADG nanoparticles. Average size of nanoparticles is 85 nm (adapted from Nimesh *et al.*, 2006 [15]).

to obtain a suspension; 2–3 μl of this suspension was deposited on a "Piranha" cleaned glass slide and allowed to dry for overnight at room temperature [18]. Subsequently, the glass surface containing the nanoparticles was imaged by AFM. The average size of nanoparticles — PPA3 alone and PPA3 loaded with DNA — was found to be 100 nm (Figure 2.2). In another study, size of the complexes of guanidinated-PEI plasmid DNA prepared at nitrogen of amine to phosphate of DNA/siRNA (N/P ratio) of 30 in phosphate buffered saline (PBS) at pH 7.4 was observed to be 350 nm [19]. Polyurethane nanoparticles observed by AFM were spherical with diameter around 209 nm for nanoparticles prepared without PEG [20]. From AFM imaging, two populations of nanoparticles were observed in the formulation prepared with PEG, 218 and 127 nm.

2.2.1.2 Dynamic light scattering (DLS)

DLS measures the temporal fluctuations of the scattered light due to the Brownian motion of the particles when a solution containing the particles is placed

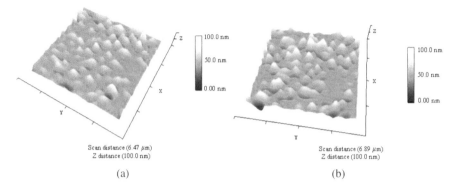

Fig. 2.2 Atomic force microscope image of (a) PPA 3 alone and (b) PPA 3–DNA complexes. Average size in each case is 100 nm. (Adapted from Nimesh *et al.*, 2007 [18]).

in the path of a monochromatic beam of light. It is also known as photon correlation spectroscopy or quasielastic light scattering. This technique gives the size in the form of hydrodynamic diameter along with the polydispersity of the sample. DLS is a non-intrusive, sensitive and powerful analytical tool and is widely used to characterize macromolecules and colloids in solution. It has also been used to measure the size and size distribution profile of nanoparticles. In a study published earlier, size of complexes was determined by DLS fitted with argon ion laser operating at 633 nm as the light source using a digital correlator [19]. Measurements were carried out at an angle perpendicular to the incident light and the data was collected over a period of 3 min. The mean particle size of the complexes of guanidinated-PEI plasmid DNA prepared at N/P ratio of 30 in PBS at pH 7.4 was observed to be 355 nm. In another study, PEI-PEG nanoparticles were characterized by DLS and TEM, and found to be in the range of 18–75 nm (hydrodynamic radii) with almost uniform population [21] (Figure 2.3).

Usually, a discrepancy in size is observed when AFM or TEM is used along with DLS to determine the size of nanoparticles; this is probably due to the two different methods used. DLS is performed on nanoparticles in water or buffer which makes them fully hydrated, whereas, AFM or TEM studies on samples dried on a glass slide or copper grid surface. Moreover, the DLS measurement presents an average size range whereas, AFM or TEM visualizes only a small number of nanoparticles. Owing to this reason, the use of different but complementary methods allows an overall evaluation of both size and

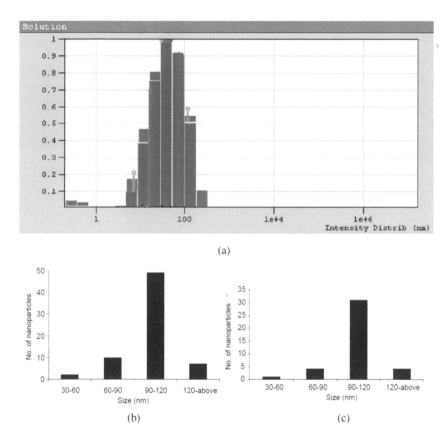

(a)

(b)　　　　　　　　　　　　　　　　(c)

Fig. 2.3　(a) Representative overlay of dynamic light scattering spectrum of PEI–PEG8000 10% (ionic crosslinked) nanoparticles in double distilled water (blue color), and 10% FCS (gray color). Size distribution obtained by TEM images of ionic nanoparticles (b) PEI–PEG6000 10% ionic alone and (c) PEI–PEG6000 10% ionic DNA loaded. Average size in each case is 100 nm. (Adapted from Nimesh *et al.*, 2006 [21]).

morphology. However, size determination under physiological conditions is an important and challenging area. The properties of nanoparticles are highly influenced by the surrounding environment. For instance, the size distribution at physiological conditions may differ from that in water or in dry state. In this regard, DLS seems to be more suitable method as it provide measurements in physiological buffers or biological fluids such as blood plasma.

2.2.2 Surface Morphology

Shape and surface morphology of nanoparticles play a key role in interaction with the cells and further internalization. TEM, AFM and Scanning Electron

Microscope (SEM) have been largely employed to determine the surface morphology of nanoparticles. SEM employs high-energy beam of electrons to scan the sample surfaces. On this instrument samples are analyzed after drying and coating with a thin layer of gold or platinum. SEM provides a direct picture of the surface of nanoparticles. Poly-D, L-lactide-co-glycolide (PLGA) nanoparticles capped with L-ascorbic acid have been analyzed by SEM for surface morphology [22]. The nanoparticles thus obtained presented a regular and smooth spherical shape, with no significant difference between nanoparticles prepared with or without ascorbic acid. Wu *et al.* analyzed (PEG-PEI)-siRNA nanoparticles at N/P ratio of 15 by SEM for morphology [23]. Here, the nanoparticles appeared spherical, uniform in size and well dispersed with average size of 240 nm. In one of our studies, surface morphology of nanoparticles prepared by crosslinking AADG with MBA was determined with TEM [15]. The nanoparticles appeared spherical with smooth surface and uniform distribution with average size of 85 nm (Figure 2.1). AFM has also been explored for surface morphology determination in number of reports. Recently, we observed nanoparticles prepared from PEI acylated with propionic anhydride (APP) with or without complexation with siRNA employing AFM (Figure 2.4) [24]. AFM investigation of these APP nanoparticles showed spherical and compact complexes with an average size of 100 nm. The three-dimensional image revealed homogeneous population with a clear absence of aggregates even

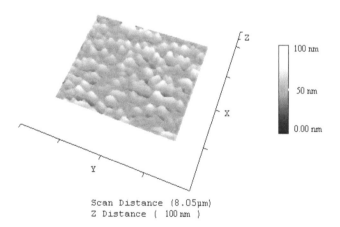

Fig. 2.4 AFM image of APP nanoparticles in double distilled water. The average particle size is 100 nm. (Adapted from Nimesh *et al.*, 2009) [24].

after 2 h. The nanoparticles maintained their morphology and shape even after binding with siRNA.

2.2.3 Surface Charge

The surface charge of nanoparticles is reported as zeta potential which is a measure of the magnitude of the repulsion or attraction between particles. When a particle is in a solution containing ions, it is surrounded by an electrical double layer of ions and counterions. The potential that exists at the hydrodynamic boundary of the particle is known as the zeta potential. It is measured by electrophoresis of the sample and determination of the velocity of the particles using laser Doppler velocimetry. The magnitude of the zeta potential relates to the potential stability of the colloidal system. On the one hand, if all the particles in suspension have a high negative or positive zeta potential, then they will tend to repel each other and there will be no tendency for aggregation. On the other hand, if the particles have low zeta potential values then there will be not be sufficient repulsive force and the particles will agglomerate. The charge of chitosan-DNA nanoparticles depend on the concentration of DNA and chitosan as well as the pH and salt content of the suspension medium. We systematically investigated the influence of salt and pH on the surface charge of chitosan/DNA nanoparticles [25]. Zeta potential measurements indicated that the particle surface charge was reduced by suspending nanoparticles in PBS at pH 6.5 (+11.4 mV) as compared to water pH 6.1 (+41.4 mV) and even became negative at pH 7.4 (−4.9 mV). This dependence of charge on pH was also observed by Mao *et al.* where electrostatically neutral particles were found in the pH range of 7.0–7.4 using a N/P ratio of 6 while the zeta potential became −20 mV at pH 8–8.5 [26].

The interaction of nanoparticles with the cells is governed by the surface charge. Polycationic polymer chitosan, interacts with the cell membranes by non-specific electrostatic forces of attraction and any receptor specific to chitosan has not been identified in cell membranes [27]. A high positive surface charge on chitosan nanoparticles is believed to allow an electrostatic interaction with the negatively charged cellular membranes leading to internalization. The results obtained in one of our studies with chitosan-DNA nanoparticles are in accordance with this theory, with higher uptake at pH 6.5 compared to pH 7.1 and 7.4, as chitosan bears higher positive at lower pH [28]. Also, significantly higher uptake of chitosan nanoparticles as compared to chitosan only was observed which could be attributed to efficient binding of condensed chitosan to cell membranes due to the presence of well defined particles at pH 6.5 in contrast to chitosan only. This dependence of uptake on the presence of condensed particles was consistent with that previously reported where chitosan nanoparticles of hydrodynamic diameter 433 ± 28 nm were reported to have higher internalization than chitosan molecules with hydrodynamic diameter of 830 ± 516 nm [29].

Functional groups on the nanoparticle's surface are the primary dictators of nanomaterial properties, such as solubility and cell surface interactions. Typically, incubation of nanomaterials with cells in media results in adsorption of serum proteins on their surface that induce the entry of nanoparticles into cells by receptor-mediated endocytosis [30]. However, in many applications, generating nanoparticles that do not interact with cell membranes or other biological macromolecules are also desirable. As an example, for *in vivo* applications, non-specific adsorption of proteins on the nanoparticle's surface can occur that may lead to particle agglomeration and clearance by RES, which hinders its ability to deliver drugs/genes to the target site. Non-specific interactions can also lead to nanoparticles binding to cell membranes and the extracellular matrix leading to inefficient tagging and detection. To avoid such issues, nanoparticles can be coated with neutral polymers such as PEG, which is well known to resist protein adoption.

2.2.4 Stability

The nanoparticles must be stable over the range of temperature and time period needed for therapeutic applications. In order to improve upon the dispersion stability of nanoparticles in liquid media, it is desirable to manipulate the particle surface by polymeric surfactants or other modifiers so as to generate an effective repulsive force between the nanoparticles. The simplest way to modify nanoparticle's surface is adsorption of polymeric dispersant. In various carboxylic acids-based anionic surfactants, such as polyacrylic acid, polyacrylic acid sodium salts and cationic surfactants, PEI is widely accepted [31–33]. However, in case of nanoparticles prepared from polycationic polymers, the high positive charge arising due to its amine groups is sufficient enough to maintain the dispersion stability. Stability while storage over a period of time is one of the important prerequisites for production of biological active molecules. It is well established that very small particles tend to aggregate among themselves to reduce the surface area, and hence to reduce the free surface energy. One of the methods to check the stability of the nanoparticles is to examine the particle size at various time points. In one of our recent publications, we reported preparation of chitosan nanoparticles by crosslinking with glutaraldehyde in reverse microemulsion where the stability was examined in terms of the particle size [34]. The lyophlized powder of nanoparticles was stored at 4°C for 1–30 days and the size of the nanoparticles was monitored by DLS. A clear suspension was obtained by sonication when lyophilized powder of nanoparticles was dispersed in water. The size of nanoparticles remained constant even after 30 days of storage, suggesting that the nanoparticles were quite stable at 4°C.

Plasmid DNA formulations prepared for transfections are unstable when stored in solution at room temperature; however, they can be stabilized by drying [35]. Lyophilization has been reported in the applications of PEI-based formulations with siRNA, oligonucleotides and ribozymes [36, 37]. It was observed that these systems retained activity when freeze-dried with glucose. It has been shown that nucleotide complexes lose transfection activity because of interactions between the complexes during the freezing step that leads to aggregation [38]. However, macromolecules such as sucrose decrease the mobility of the particles thereby decreasing the interactions between them by steric hindrance. Anderson *et al.* also observed a requirement for a lyoprotectant in their freeze-dried chitosan system to prevent aggregation and loss of silencing activity [39].

2.2.5   Sterility

Sterility is highly desirable for both *in vitro* and *in vivo* application of nanoparticles for drug and gene delivery. For clinical use, parenteral drug delivery systems have to meet the pharmacopoeial requirements of sterility. The chemical or physical liability of the polymer matrix usually limits most conventional methods for obtaining acceptable sterile products [40]. In the case of polymer nanoparticles used as drug carriers, it is a challenge to find a satisfactory sterilization technique which would be able to keep the supramolecular and molecular structure of the colloids intact. With chemical sterilization by gases such as ethylene oxide, toxicological problems may be encountered due to toxic residues. Numerous studies have shown the effects of γ-irradiation on the stability and the safety of colloidal carriers based on polyesters, principally microparticles and nanoparticles [41, 42]. Therefore, the selection of a suitable sterilization method for such type of formulations is crucial to ensure their physical and chemical integrity, performance and safety *in vivo*.

As an alternative technique, sterile filtration based on physical removal of contained micro-organisms through 0.22 μm membrane filters may be considered as the appropriate method for chemically or thermally sensitive materials since it has no adverse effect on the polymer and the drug [43–45]. Moreover, it neither adversely affects the drug release properties and the stability of a formulation nor the chemical stability of ingredients. Nevertheless, this sterilization method can only be used for nanoparticles with a mean size significantly below membrane cutoff and with a narrow size distribution to avoid membrane clogging. However, this technique is not suitable for larger nanoparticles (exceeding 200 nm) when the drug is adsorbed at the nanoparticles' surface or when the colloidal suspensions are too viscous [46, 47]. In one of our studies, we sterilized chitosan solutions by the sterile filtration technique before complexation with DNA, followed by incubation and transfection of complexes in sterile conditions [25]. The complexes were efficient without any signs of contamination i.e. no bacterial or mycoplasma contamination.

Heat sterilization by autoclaving is a highly effective technique involving high temperatures, which may influence decomposition or degradation of active ingredient as well as the nanoparticle material, i.e., polymer [41, 48]. A significant increase in particle size was reported after autoclaving of the unloaded polybutylcyanoacrylate nanoparticle suspensions and the nanoparticle powders were characterized by impaired resuspension characteristics. These were attributed to the swelling of polymeric membrane. On the one hand, sterilization by autoclaving induces the degradation of polyesters by hydrolysis and these polymers are also heat sensitive due to their thermoplastic nature. On the other hand, solid lipid nanoparticles can be sterilized by autoclaving, maintaining an almost spherical shape, without any significant increase in size or distribution [49, 50]. The Food and Drug Administration requires that sterile pharmaceutical products be free of viable micro-organisms. Sterility testing of pharmaceutical products provides added assurance that the product is sterile. Sterility testing is typically done by inoculating the drug product into microbial growth media including gram-positive and gram-negative bacteria, spore forming bacteria, yeast and fungi followed by visual inspection for growth during incubation for a specified time period. A lack of visual growth indicates that the drug product samples tested were sterile.

References

[1] Decuzzi, P., R. Pasqualini, W. Arap, and M. Ferrari. Intravascular delivery of particulate systems: Does geometry really matter? *Pharmaceutical Research*, 26: 235–243, 2009.

[2] Patil, Y.B., U.S. Toti, A. Khdair, L. Ma, and J. Panyam. Single-step surface functionalization of polymeric nanoparticles for targeted drug delivery. *Biomaterials*, 30: 859–866, 2009.

[3] Birrenbach, G. and P.P. Speiser. Polymerized micelles and their use as adjuvants in immunology. *Journal of Pharmaceutical Sciences*, 65: 1763–1766, 1976.

[4] Kreuter, J. and P.P. Speiser. *In vitro* studies of poly(methyl methacrylate) adjuvants. *Journal of Pharmaceutical Sciences*, 65: 1624–1627, 1976.

[5] Soppimath, K.S., T.M. Aminabhavi, A.R. Kulkarni, and W.E. Rudzinski. Biodegradable polymeric nanoparticles as drug delivery devices. *Journal of Controlled Release*, 70: 1–20, 2001.

[6] Ogris, M. and E. Wagner. Targeting tumors with nonviral gene delivery systems. *Drug discovery today*, 7: 479–485, 2002.

[7] Dauty, E., J.S. Remy, T. Blessing, and J.P. Behr. Dimerizable cationic detergents with a low cmc condense plasmid DNA into nanometric particles and transfect cells in culture. *Journal of the American Chemical Society*, 123: 9227–9234, 2001.

[8] Lee, H., S.K. Williams, S.D. Allison, and T.J. Anchordoquy. Analysis of self-assembled cationic lipid-DNA gene carrier complexes using flow field-flow fractionation and light scattering. *Analytical Chemistry*, 73: 837–843, 2001.

[9] Sakurai, F., R. Inoue, Y. Nishino, A. Okuda, O. Matsumoto, T. Taga, F. Yamashita, Y. Takakura, and M. Hashida. Effect of DNA/liposome mixing ratio on the physicochemical characteristics, cellular uptake and intracellular trafficking of plasmid DNA/cationic liposome complexes and subsequent gene expression. *Journal of Controlled Release*, 66: 255–269, 2000.

[10] Desai, M.P., V. Labhasetwar, E. Walter, R.J. Levy, and G.L. Amidon. The mechanism of uptake of biodegradable microparticles in Caco-2 cells is size dependent. *Pharmaceutical research*, 14: 1568–1573, 1997.

[11] Zauner, W., N.A. Farrow, and A.M. Haines. *In vitro* uptake of polystyrene microspheres: Effect of particle size, cell line and cell density. *Journal of Controlled Release*, 71: 39–51, 2001.

[12] Prabha, S., W.Z. Zhou, J. Panyam, and V. Labhasetwar. Size-dependency of nanoparticle-mediated gene transfection: Studies with fractionated nanoparticles. *International journal of pharmaceutics*, 244: 105–115, 2002.

[13] Chithrani, B.D., A.A. Ghazani, and W.C. Chan. Determining the size and shape dependence of gold nanoparticle uptake into mammalian cells. *Nano Letters*, 6: 662–668, 2006.

[14] Chithrani, B.D. and W.C. Chan. Elucidating the mechanism of cellular uptake and removal of protein-coated gold nanoparticles of different sizes and shapes. *Nano Letters*, 7: 1542–1550, 2007.

[15] Nimesh, S., R. Manchanda, R. Kumar, A. Saxena, P. Chaudhary, V. Yadav, S. Mozumdar, and R. Chandra. Preparation, characterization and *in vitro* drug release studies of novel polymeric nanoparticles. *International Journal of Pharmaceutics*, 323: 146–152, 2006.

[16] Patnaik, S., A.K. Sharma, B.S. Garg, R.P. Gandhi, and K.C. Gupta. Photoregulation of drug release in azo-dextran nanogels. *International journal of Pharmaceutics*, 342: 184–193, 2007.

[17] Montasser, I., H. Fessi, and A.W. Coleman. Atomic force microscopy imaging of novel type of polymeric colloidal nanostructures. *European Journal of Pharmaceutics and Biopharmaceutics*, 54: 281–284, 2002.

[18] Nimesh, S., A. Aggarwal, P. Kumar, Y. Singh, K.C. Gupta, and R. Chandra. Influence of acyl chain length on transfection mediated by acylated PEI nanoparticles. *International Journal of Pharmaceutics*, 337: 265–274, 2007.

[19] Nimesh, S. and R. Chandra. Guanidinium-grafted polyethylenimine: An efficient transfecting agent for mammalian cells. *European Journal of Pharmaceutics and Biopharmaceutics*, 68: 647–655, 2008.

[20] Zanetti-Ramos, B.G., M.B. Fritzen-Garcia, C.S. De Oliveira, A.A. Pasa, V. Soldi, R. Borsali, and T.B. Creczynski-Pasa. Dynamic light scattering and atomic force microscopy techniques for size determination of polyurethane nanoparticles. *Materials Science and Engineering*: C, 29: 638–640, 2009.

[21] Nimesh, S., A. Goyal, V. Pawar, S. Jayaraman, P. Kumar, R. Chandra, Y. Singh, and K.C. Gupta. Polyethylenimine nanoparticles as efficient transfecting agents for mammalian cells. *Journal of Controlled Release*, 110: 457–468, 2006.

[22] Martins, D., L. Frungillo, M.C. Anazzetti, P.S. Melo, and N. Duran. Antitumoral activity of L-ascorbic acid-poly- D,L-(lactide-co-glycolide) nanoparticles containing violacein. *International Journal of Nanomedicine*, 5: 77–85.

[23] Wu, Y., W. Wang, Y. Chen, K. Huang, X. Shuai, Q. Chen, X. Li, and G. Lian. The investigation of polymer-siRNA nanoparticle for gene therapy of gastric cancer *in vitro*. *International Journal of Nanomedicine*, 5:129–136.

[24] Nimesh, S. and R. Chandra. Polyethylenimine nanoparticles as an efficient *in vitro* siRNA delivery system. *European Journal of Pharmaceutics and Biopharmaceutics*, 2009.

[25] Nimesh, S., M.M. Thibault, M. Lavertu, and M.D. Buschmann. Enhanced gene delivery mediated by low molecular weight chitosan/DNA complexes: Effect of pH and serum. *Molecular Biotechnology*, 46: 182–196, 2010.

[26] Mao, H.Q., K. Roy, V.L. Troung-Le, K.A. Janes, K.Y. Lin, Y. Wang, J.T. August, and K.W. Leong. Chitosan-DNA nanoparticles as gene carriers: Synthesis, characterization and transfection efficiency. *Journal of Controlled Release*, 70: 399–421, 2001.

[27] Schipper, N.G., S. Olsson, J.A. Hoogstraate, A.G. deBoer, K.M. Varum, and P. Artursson. Chitosans as absorption enhancers for poorly absorbable drugs 2: Mechanism of absorption enhancement. *Pharmaceutical Research*, 14: 923–929, 1997.

[28] Nimesh, S., M. Thibault, M. Lavertu, and M. Buschmann. Enhanced gene delivery mediated by low molecular weight chitosan/DNA complexes: Effect of pH and serum. *Molecular Biotechnology*, 46(2): 182–196.

[29] Ma, Z. and L.Y. Lim. Uptake of chitosan and associated insulin in caco-2 cell monolayers: A comparison between chitosan molecules and chitosan nanoparticles. *Pharmaceutical Research*, 20: 1812–1819, 2003.

[30] Khan, J.A., B. Pillai, T.K. Das, Y. Singh, and S. Maiti. Molecular effects of uptake of gold nanoparticles in Hela cells. *Chemistry and Biochemistry*, 8: 1237–1240, 2007.

[31] Prabhakaran, K., C.S. Kumbhar, S. Raghunath, N.M. Gokhale, and S.C. Sharma. Effect of concentration of ammonium poly(acrylate) dispersant and MgO on coagulation characteristics of aqueous alumina direct coagulation casting slurries. *Journal of the American Ceramic Society*, 91: 1933–1938, 2008.

[32] Sato, K., S. Kondo, M. Tsukada, T. Ishigaki, and H.O. Kamiya. Influence of solid fraction on the optimum molecular weight of polymer dispersants in aqueous TiO_2 nanoparticle suspensions. *Journal of the American Ceramic Society*, 90: 3401–3406, 2007.

[33] Laarz, E., A. Meurk, J.A. Yanez, and L. Bergström. Silicon itride olloidal probe measurements: Interparticle forces and the role of surface-segment interactions in poly(acrylic acid) adsorption from aqueous solution. *Journal of the American Ceramic Society*, 84: 1675–1682, 2001.

[34] Manchanda, R. and S. Nimesh. Controlled size chitosan nanoparticles as an efficient, biocompatible oligonucleotides delivery system. *Journal of Applied Polymer Science*, 118: 2071–2077, 2010.

[35] Romoren, K., A. Aaberge, G. Smistad, B.J. Thu, and O. Evensen. Long-term stability of chitosan-based polyplexes. *Pharmaceutical Research*, 21: 2340–2346, 2004.

[36] Werth, S., B. Urban-Klein, L. Dai, S. Hobel, M. Grzelinski, U. Bakowsky, F. Czubayko, and A. Aigner. A low molecular weight fraction of polyethylenimine (PEI) displays increased transfection efficiency of DNA and siRNA in fresh or lyophilized complexes. *Journal of Controlled Release*, 112: 257–270, 2006.

[37] Brus, C., E. Kleemann, A. Aigner, F. Czubayko, and T. Kissel. Stabilization of oligonucleotide-polyethylenimine complexes by freeze-drying: Physicochemical and biological characterization. *Journal of Controlled Release*, 95: 119–131, 2004.

[38] Allison, S.D., M.C. Molina, and T.J. Anchordoquy. Stabilization of lipid/DNA complexes during the freezing step of the lyophilization process: The particle isolation hypothesis. *Biochimica et Biophysica Acta*, 1468: 127–138, 2000.

[39] Andersen, M.O., K.A. Howard, S.R. Paludan, F. Besenbacher, and J. Kjems. Delivery of siRNA from lyophilized polymeric surfaces. *Biomaterials*, 29: 506–512, 2008.

[40] Athanasiou, K.A., G.G. Niederauer, and C.M. Agrawal. Sterilization, toxicity, biocompatibility and clinical applications of polylactic acid/polyglycolic acid copolymers. *Biomaterials*, 17: 93–102, 1996.

[41] Memisoglu-Bilensoyand, E. and A.A. Hincal. Sterile, injectable cyclodextrin nanoparticles: Effects of gamma irradiation and autoclaving. *International Journal of Pharmaceutics*, 311: 203–208, 2006.

[42] Volland, C., M. Wolff, and T. Kissel. The influence of terminal gamma-sterilization on captopril containing poly(d,l-lactide-co-glycolide) microspheres. *Journal of Controlled Release*, 31: 293–305, 1994.

[43] Goldbach, P., H. Brochart, P. Wehrlé, and A. Stamm. Sterile filtration of liposomes: Retention of encapsulated carboxyfluorescein. *International Journal of Pharmaceutics*, 117: 225–230, 1995.

[44] Konan, Y.N., R. Gurny, and E. Allémann. Preparation and characterization of sterile and freeze-dried sub-200 nm nanoparticles. *International Journal of Pharmaceutics*, 233: 239–252, 2002.

[45] Maksimenko, O., E. Pavlov, E. Toushov, A. Molin, Y. Stukalov, T. Prudskova, V. Feldman, J. Kreuter, and S. Gelperina. Radiation sterilisation of doxorubicin bound to poly(butyl cyanoacrylate) nanoparticles. *International Journal of Pharmaceutics*, 356: 325–332, 2008.

[46] Brigger, I., L. Armand-Lefevre, P. Chaminade, M. Besnard, Y. Rigaldie, A. Largeteau, A. Andremont, L. Grislain, G. Demazeau, and P. Couvreur. The stenlying effect of high hydrostatic pressure on thermally and hydrolytically labile nanosized carriers. *Pharmaceutical Research*, 20: 674–683, 2003.

[47] Tsukada, Y., K. Hara, Y. Bando, C.C. Huang, Y. Kousaka, Y. Kawashima, R. Morishita, and H. Tsujimoto. Particle size control of poly(dl-lactide-co-glycolide) nanospheres for sterile applications. *International Journal of Pharmaceutics*, 370: 196–201, 2009.

[48] Wörle, G., B. Siekmann, M.H.J. Koch, and H. Bunjes. Transformation of vesicular into cubic nanoparticles by autoclaving of aqueous monoolein/poloxamer dispersions. *European Journal of Pharmaceutical Sciences*, 27: 44–53, 2006.

[49] Cavalli, R., O. Caputo, M.E. Carlotti, M. Trotta, C. Scarnecchia, and M.R. Gasco. Sterilization and freeze–drying of drug-free and drug-loaded solid lipid nanoparticles. *International Journal of Pharmaceutics*, 148: 47–54, 1997.

[50] Sanna, V., N. Kirschvink, P. Gustin, E. Gavini, I. Roland, L. Delattre, and B. Evrard. Preparation and *in vivo* toxicity study of solid lipid microparticles as carrier for pulmonary administration. *AAPS PharmSciTech*, 5: e27, 2004.

3

In Vitro and *In Vivo* Characterization of Nanoparticles

3.1 Introduction

Interaction of nanoparticles with the cells is a prerequisite for efficient gene and drug delivery. Understanding the biophysical interaction of nanoparticles with cell membranes is critical for developing an effective nanocarrier system for therapeutic applications. The nanoparticles have been thought to interact with the plasma membrane of the cells through electrostatic forces. The plasma membrane has been observed to be negatively charged due to presence of various proteins and interacts efficiently with the positively charged nanoparticles. Hence, positive charge over nanoparticles' surface is highly advantageous, and, cationic polymers which easily impart positive charge to the nanoparticles are preferred candidates. However, this positive charge in access is also toxic for the cells. The interaction of nanoparticles followed by its delivery activity is a crucial and multistep process that needs to be well characterized in order to develop vectors with desirable qualities.

3.2 *In Vitro* Characterization of Nanoparticles

3.2.1 Binding and Uptake

The first and one of the most crucial steps towards application of nanoparticles is its binding with the cell membranes followed by uptake. Nanoparticles with net positive charge and size less than a micron interact with the cell surface proteins and sugars of the plasma membrane, usually negatively charged,

S. Nimesh and R. Chandra,
Theory, Techniques and Applications of Nanotechnology in Gene Silencing, 27–46.
© 2011 *River Publishers. All rights reserved.*

to gain an anchor on the cell surface. Once attached to the cell surface by non-specific means, their presence stimulates the endocytic pathway leading to formation of endosomal vesicles by a process known as endocytosis. The life cycle of endosomes is strictly defined by various extra endosomal factors and cytoskeleton. Endocytosis is a highly regulated multistep process. First, the material is engulfed in membrane invaginations which are pinched off to form membrane bound vesicles, also known as endosomes (or phagosomes in case of phagocytosis). Cells contain heterogeneous populations of endosomes equipped with distinct endocytic machinery, which originate at different sites of the cell membrane. Second, the endosomes deliver the material to various specialized vesicular structures, which enables sorting of materials towards different destinations. And finally, the material is delivered to various intracellular compartments, recycled to the extracellular milieu or delivered across cells (a process known as "transcytosis" in polarized cells). Generally, endocytosis can be further categorized into two broad classes — phagocytosis (the uptake of large particles) and pinocytosis (the uptake of fluids and solutes).

This interaction of nanoparticles with the cells has been well investigated by sophisticated techniques such as flow cytometry and confocal microscopy. In one of our studies, flow cytometry was employed to investigate the uptake of chitosan/DNA nanoparticles by HEK 293 cells [1]. Flow cytometry was used to quantify the uptake of nanoparticles by the cells, however, it does not generally discriminate between membrane-bound and internalized fluorescent moieties, so additional procedures were included to minimize any contribution of surface-bound particles in measuring uptake. A simple and efficient method involving treatment of the cells with trypsin followed by several washes before fluorescence assisted cell sorting (FACS) analysis was used to remove surface bound complexes, and assess cellular uptake of fluorescent chitosan and complexes. Uptake of complexes by HEK 293 cells was pH and serum dependent, with maximum uptake occurring in medium at pH 6.5 supplemented with 10% FBS. The uptake of nanoparticles as well as of chitosan alone was higher in the presence of serum at all the three investigated pHs [1]. Chitosan interacts with the cell membranes by non-specific electrostatic forces of attraction and any receptor specific to chitosan has not been identified in cell membranes [2]. A high positive surface charge on chitosan nanoparticles is thought to allow an electrostatic interaction with the negatively charged cellular membranes leading to internalization. Further, to fully substantiate the uptake results from flow cytometry analysis, we assessed the cellular internalization of nanoparticles with confocal microscopy [1]. Although this technique is more qualitative than quantitative, it provides a direct observation of the localization of the fluorescent nanoparticles in the cells providing evidence of cellular internalization. During cellular tracking of the fluorescent chitosan/DNA nanoparticles, fluorescence appeared to be distributed throughout inside the cells at pH 6.5 with serum (Figure 3.1). However, cells incubated with the complexes at pH 7.1 and 7.4 with serum showed large aggregates found near the cell membranes with only small amounts that were internalized [1].

We observed strong correlation between cell binding (Figure 3.2a) and cell uptake (Figure 3.2b), with a notable increase in uptake for the two chitosans with high degree of deacetylation (DDA) (92%) compared to lower DDA chitosans [3]. The correlation between binding and uptake underscores the necessity of establishing cell contact with polyplexes. The number of cells positive for uptake (Figure 3.2c) peaked at around 8 hours when nearly 100% of the cells contained internalized polyplexes. In contrast, the total amount taken up per cell increased continuously throughout time until the media was changed, indicating that no saturation of internalization occurs at any time point. The process of internalization of nanoparticles has been reported to depend on the parameters discussed in the following sections.

3.2.1.1 Particle charge

Cellular internalization depends on the surface charge of nanoparticles. Presently, majority of reports suggest that the positively charged nanoparticles predominantly internalize through clathrin-mediated endocytosis with some fraction utilizing macropinocytosis. Examples include cationic nanoparticles of totally different origin — stearylamine-coated PEG-*co*-PLA, PLGA modified with poly-L-lysine (PLL), chitosan, etc [4, 5]. On the one hand, PEI based polyplexes are strongly cationic and yet may utilize multiple pathways including caveolae-mediated endocytosis [6]. On the other hand, negatively charged nanoparticles, such as DOXIL®, micelles and quantum dots are more likely to utilize caveolae-mediated endocytosis [7, 8]. Since cell membranes are generally negatively charged, it is widely accepted that negatively charged nanoparticles should internalize slower compared to their positively charged

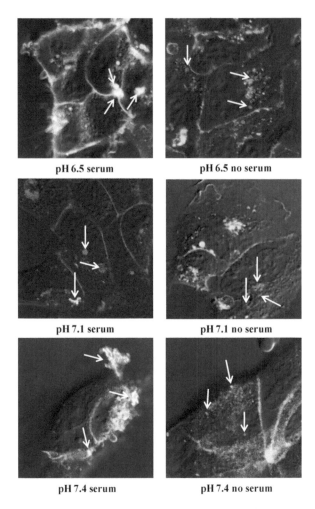

pH 6.5 serum   pH 6.5 no serum

pH 7.1 serum   pH 7.1 no serum

pH 7.4 serum   pH 7.4 no serum

Fig. 3.1   Confocal microscope images of HEK 293 cells transfected at different pH. Cells transfected with chitosan/DNA complexes, where chitosan labeled with rhodamine and DNA with fluorescein, visualized under confocal microscope 24 h post-transfection. Blue color — Staining of cell membrane, Red color — Rhodamine labelled chitosan only, Green color — Fluorescein labeled DNA only and Yellow-Colocalization showing Chitosan/DNA complex. At pH 6.5 with serum, a large amount of complexes are visible inside the cell (white arrows) whereas at pH 6.5 without serum, a large amount of free chitosan can be seen inside the cell (white arrows). At pH 7.1 with serum, a small amount of chitosan and complexes were visible (white arrows) whereas at pH 7.1 without serum only a small amount of free chitosan was seen inside cells (white arrows) without any internalization of complexes. At pH 7.4 with serum, large aggregates were observed outside the cells external to the cell membrane (white arrows) while at pH 7.4 without serum, a small amount of complexes were observed on cell membranes (white arrows). Adapted from Nimesh *et al.*, 2010 [1].

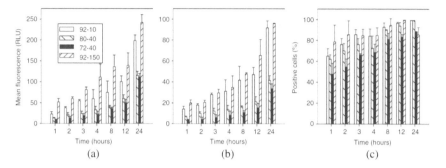

Fig. 3.2 Kinetics of cellular binding and uptake of polyplexes prepared with chitosans of different DDA and molecular weight (MW). HEK 293 cells were incubated with fluorescent chitosan polyplexes for the indicated periods of time and analyzed by flow cytometry. Flow cytometry quantitative analysis of mean fluorescence per cell for polyplex (a) binding and (b) uptake and (c) of % cells with internalized polyplexes was performed following trypsinization and extensive washes, except for (a) cell binding, where cells were detached by enzyme-free cell dissociation buffer and analyzed directly. (b) Mean uptake levels per cell and (c) % positive cells were obtained from the same set of flow cytometry data. Graphs show that binding and uptake are time and DDA-dependent, with both 92% DDA chitosans binding more effectively than the lower DDA chitosans, resulting in increased uptake. Results are the average of three ($N = 3$) independent experiments $\pm$ SD where each experiment included two replicates. RLU, relative light units. Adapted from Thibault *et al.*, 2010 [3].

counterparts. However, it is unclear from literature whether neutral nanoparticles show any preference for specific cellular entry routes.

3.2.1.2 Particle size

The size of nanoparticles has been thought to play a dominant role in their uptake within different endocytic vesicles that greatly vary in size. Hence, the requirement to keep the particle small between 10 to 100 nm, to enter the endocytic vesicles was postulated. However, while the small size may favor rapid entry into cells, there is no cut-off limit for sizes upto at least 5 μm to gain cellular entry of some materials through pinocytosis. The largest particles are more likely to enter cells through macropinocytosis [9]. For example, the negatively charged polystyrene nanoparticles showed clathrin-mediated endocytosis based entry for 43 nm particles and caveolae-mediated entry for 25 nm particles [10]. High polydispersity in the size of nanoparticles appears to be a major problem in identifying the effects of the size. Moreover, in some cases, such as highly polydisperse chitosan nanoparticles, effect of the size of the particles is less pronounced in defining the entry pathway than the chemical composition of the nanomaterial [11].

3.2.1.3 Particle shape

A study by Champion *et al.* described the effect of the shape of particles on phagocytosis in alveolar rat macrophages [12]. They prepared polystyrene-based particles of more than twenty shapes including spheres, rectangles, rods, worms, oblate ellipses, elliptical disks and UFO-like. The particle sizes were primarily 1 to about 10 μm. However, the size was not a rate limiting factor in phagocytosis [13]. All particle shapes independent of their sizes were capable of initiating phagocytosis in at least one orientation. However, unexpectedly, the crucial role in phagocytosis was played by the local particle shape at the point of attachment of particle to the macrophage [12]. For example, a macrophage attached to a sharper side of the ellipse would internalize this ellipse in a few minutes. In contrast, a macrophage attached to a dull side would not internalize the same ellipse for many hours. Spheres were internalized from any point of attachment independent of their size. Although, the particle size played a much lesser role in the initiation of the phagocytosis, it could of course affect its completion especially when the particle volume exceeded that of a cell. The effect of geometry in phagocytosis was quantified by measuring the angle between the membrane normal at the point of initial contact and the line defining the particle curvature at this point [13]. If this angle exceeded some critical value (45°) the macrophages would lose the ability to entrap particles and attach to these particles in a process similar to spreading. The authors suggested that the shape of the particle at the point of attachment would define the complexity of actin structures needed to be rearranged to allow engulfment. Above the critical point the necessary actin structures could not be created and the macrophages would switch to the spreading behavior. Altogether, this eloquent study clearly demonstrated the relationship between cellular transport by phagocytosis and physical properties of the phagocytosed materials.

3.2.1.4 Cell Type

It appears that the relationship between the nanoparticle's cellular trafficking and cell type is least explored. Few examples report that the cell type may be critical in defining the nanoparticle entry and final destination in the cells. It is noteworthy that the differential endocytic pathways in normal and tumor cells may be a gateway for selective targeting of novel nanomaterials into tumors [7]. Most studies concerning nanoparticle's trafficking did not emphasize the relationship between cell origin and availability of various endocytic pathways present in these cells. Also, the known cellular pathways may be differentially presented or even totally absent in selected cell types depending on the cell phenotype, and in some cases even growing conditions, such as cell density, presence of growth factors, etc. Obviously, an indepth understanding of cell biology and it's relation to nanomaterial science is most critical for advancement in the area of nanomedicine and drug delivery.

3.2.2 Cytotoxicity

An important challenge for the success of nanomedicine-based therapies is the need for an acceptable and efficient delivery system that has minimal toxicity and maximum patient safety. Nanoparticles intended for imaging and drug delivery are often purposely coated with bioconjugates such as DNA, proteins, and monoclonal antibodies to target specific cells. As these nanoparticles are engineered to interact with cells, it is important to ensure that these enhancements do not cause any adverse effects; more significantly, whether naked or coated nanoparticles will undergo biodegradation in the cellular environment and what cellular responses these degraded nanoparticles will induce. A plethora of reports have been published reporting *in vitro* cytotoxicity studies of nanoparticles using different cell lines, incubation times, and colorimetric assays. These studies employ a wide range of nanoparticle concentrations and exposure times, making it difficult to determine whether the cytotoxicity observed is physiologically relevant. Additionally, different groups choose to use various cell lines as well as culturing conditions, which render direct comparisons between the available studies difficult.

The initial step toward understanding how an agent will react in the body often involves studying the cell cultures. As compared to animal studies, cellular testing is less ethically ambiguous, is easier to control and reproduce, and less expensive. While studying cytotoxicity, it is noteworthy that cell cultures are sensitive to changes in their environment such as fluctuations in temperature, pH, nutrient and waste concentrations, in addition to the concentration of potentially toxic agent being tested. Hence, monitoring of experimental conditions is crucial to ensure that the measured cell death corresponds to the toxicity of the added nanoparticles versus the unstable culturing conditions.

Additionally, nanoparticles can also adsorb dyes and be redox active for which it is important that the appropriate cytotoxicity assay is chosen. Conducting multiple tests is advantageous to ensure valid conclusions are drawn.

One of the simplest cytotoxicity tests involves visual inspection of the cells with bright field microscopy for changes in cellular or nuclear morphology. Using this technique, we observed that PEI polymer induced considerable toxicity and cell morbidity in COS-1 cells as compared to PEI-PEG nanoparticles [14]. In another study, we used cell confluency, a qualitative measure of cell viability based upon cell coverage on the well surface, and judged by microscopy, as an indication of toxicity of chitosan/DNA nanoparticles used for transfection as compared to Lipofectamine [1]. The cells were more confluent and maintained their normal shape when transfected with chitosan/DNA nanoparticles in contrast to Lipofectamine. Exposure to certain cytotoxic agents can compromise the cell membrane, which allows cellular contents to leak out. Viability tests based on this include the neutral red and trypan blue assays. Neutral red or toluylene red is a weak cationic dye that can cross the plasma membrane by diffusion. This dye tends to accumulate in lysosomes within the cell. If the cell membrane is altered, the uptake of neutral red is decreased and can leak out, allowing for discernment between live and dead cells. Cytotoxicity can be quantified by taking spectrophotonic measurements of the neutral red uptake under varying exposure conditions. Studies by Flahaut *et al.* and Monteiro-Riviere *et al.* determining the cytotoxicity of carbon nanotubes utilized the neutral red assay [15, 16]. Trypan blue, a diazo dye, is only permeable to cells with compromised membranes; therefore, dead cells are stained blue while live cells remain colorless. The amount of cell death can be determined via light microscopy [17]. This assay was used by Bottini *et al.* and Goodman *et al.* to determine the cytotoxicity of single walled nanotubes and gold nanoparticles [18, 19].

Along with distinguishing between live and dead cells by detecting damaged cell membranes, other colorimetric cytotoxicity assays aimed at detecting the metabolic activity of the cells have been employed. Mitochondrial activity of cells has been tested using tetrazolium salts as mitochondrial dehydrogenase enzymes cleave the tetrazolium ring. Only active mitochondria contain these enzymes; therefore, the reaction only occurs in living cells [20]. The most widely employed assay is the 3-(4,5-dimethylthiazol-2-yl)-2,5-diphenyl tetrazolium bromide (MTT), which is pale yellow in solution but produces a dark-blue formazan product within live cells. We also utilized MTT for determination of cytotoxicity of nanoparticles prepared by derivatizing PEI [14, 21–24]. Recently, the toxicity of nanoparticles was evaluated by colorimetric Alamar blue assay [25, 26]. The blue colored reagent Alamar blue contains resazurin which is reduced to a pink coloured resorufin by the metabolic mitochondrial activity of viable cells and can be quantified colorimetrically and fluorimetrically. We used Alamar blue to assess potential cytotoxicity of nanoparticles prepared from high DDA (92%) and low MW (10 kDa) chitosan [1]. The chitosan/DNA nanoparticles were found to be only slightly toxic, where after 48 h of incubation more than 85% of cells were viable at pH 6.5 and 96% at pH 7.1 (Figure 3.3) [1]. During application of nanoparticles, the major biological effects involve interactions with cellular components such as the plasma membranes, organelles, or macromolecules. As different nanoparticles behave differently, they can trigger distinctive biological responses, therefore, it is important that cytotoxicity studies are conducted for each nanoparticle type.

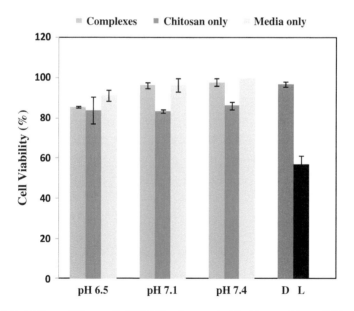

Fig. 3.3 Cell viability at different pH. HEK 293 cells treated with chitosan, chitosan/DNA complexes at different pH, 48 h post-transfection cell viability determined using Alamar blue. Cells at pH 7.4 in complete media taken as 100% viable, other controls include D: Cells incubated with DNA alone, L: Cells transfected using Lipofectamine. Values are mean ± SD, $n = 3$. Adapted from Nimesh *et al.*, 2010 [1].

3.3 *In Vivo* Characterization

Nanoparticles owing to their unique properties are rapidly emerging as promising candidates for a range of applications in nanomedicine, ranging from *in vitro* diagnostic assays to *in vivo* localized imaging, drug delivery and therapy [27–29]. Recently, multifunctional nanoparticles with diagnostic and therapeutic functions ("theragnostics") have been proposed for *in vivo* applications, prompting an ongoing demand to understand their *in vivo* biodistribution and potential clearance mechanisms [30]. The efficacy and safety of nanoparticles depends on the biodistribution, which in turn dictates the clinical relevance of nanoparticle-based technology. For instance, a clear understanding of the extent to which nanoparticles localize at a site of interest is important for evaluating how effectively a therapeutic nanoparticle is passively or actively targeted. Evaluation of potential unwanted accumulation in other tissues and characterization of the long-term fate of the nanoparticles (indefinite residence, metabolism or excretion) are required for understanding the safety of administered nanoparticles [31]. Realizing the importance, the biodistribution of nanoparticles has been studied since the development of the first nanoparticles for biomedical applications, with a large portion of studies conducted on polymeric nanoparticles [32].

Nanoparticles can be employed to render protection or reduce renal clearance for easily degraded or short half-life drugs, such as small peptides and nucleic acids, for a prolonged therapeutic effect. However, nanoparticles may also improve the availability of drugs to certain tissues and thus, cause new side effects. The pharmacokinetic profiles of the parent drug and the drug entrapped in the nanoparticles are often different. Hence, monitoring of the pharmacokinetics and biodistribution of nanoparticles is essential to understand and predict their efficacy and side effects. The pharmacokinetics profile of the nanoparticles depends on their chemical and physical properties, such as size, charge, and surface chemistry. Pharmacokinetics study involves measuring drug concentrations in all major tissues after drug administration over a period of time until the elimination phase. It is necessary to monitor the drug concentration long enough to fully describe the behavior of the drug or nanoparticles *in vivo* (usually 3 × half-life).

Upon intravenous administration of nanoparticles, a variety of serum proteins bind to the surface of the nanoparticles, which are recognized by the scavenger receptor on the macrophage cell surface and internalized, leading to a significant removal of nanoparticles from the circulation. The serum proteins binding on the nanoparticles are also termed opsonins, and the process as opsonization. The macrophages responsible for removal of such nanoparticles are known as the RES or mononuclear phagocyte system (MPS). Minimizing protein interaction and binding to nanoparticles is one of the major factors for developing a long circulation nanoparticle formulation.

3.3.1 Protein Binding to Nanoparticles

Nanoparticles are exposed to a rich environment of proteins, cells and tissues upon exposure to bloodstream. Probably, the proteins at high concentrations in plasma and with high association rates will initially occupy the surface of the nanoparticle. However, over time these proteins may dissociate and replaced by proteins of lower concentrations, slower exchange rates, and/or higher affinities [2, 6]. The whole process of competitive adsorption of proteins onto a limited surface based on abundance, affinities, and incubation time is collectively known as the "Vroman Effect" [24, 61, 62]. This effect significantly contributes to particle distribution throughout the body. As the particles distribute from the blood to various locations, the differences in protein levels, as well as their affinities for binding, may play a part in determining how the protein corona evolves as the nanoparticle moves from one compartment to another. The kinetics of protein adsorption can be studied in several ways. A study on 50 nm lecithin-coated polystyrene nanospheres used a combination of sodium dodecyl sulfate-polyacrylamide gel electrophoresis (SDS-PAGE) and western blotting to analyze samples at various time points [58]. This study revealed a quantitative and qualitative profile of serum proteins adsorbed on the surface of the nanoparticles over periods from 5 min to 360 min. SDS-PAGE and western blotting provide vital information about the proteins adsorbed onto the nanoparticles. However, it is very difficult to obtain quantitative results from these gel methods as they are mainly used for comparison purposes. SDS-PAGE can detect anywhere between 1 and 50 ng of a single protein band, depending on the stain used, while western blot limitations are dependent on the antibodies used as well as the conjugated substrate of choice. All these conditions would have to be optimized for the particular protein of interest. Another study evaluated the kinetics of protein binding to PEG–polyhexadeclycyanoacrylate (PHDCA) nanoparticles using a system that separates fluorescent dye labeled protein-SDS complexes electrokinetically in a gel media, and separates the proteins based on molecular weight [9]. A few established techniques have recently been used to analyze the kinetic properties of protein binding to nanoparticle surfaces in a less potentially disruptive manner than centrifugation. These techniques include size-exclusion chromatography, isothermal titration calorimetry (ITC), and surface plasmon resonance [2, 40]. In various N-isopropylacrylamide/N-tert-butylacrylamide (NIPAM/BAM) copolymer nanoparticles, isothermal titration calorimetry was used to determine the stoichiometry and affinity of human serum albumin (HSA) to the nanoparticle surface.

3.3.2   Factors Affecting Protein Binding to Nanoparticles

Most of the studies investigating the effect of protein binding on uptake have been conducted by either preincubating particles with serum/plasma or by preincubating particles with individual proteins or attaching individual proteins to the surface of the particles, and evaluating uptake by macrophages. On the basis of published work, it can be deduced that neutral particles have a comparatively slower opsonization rate than charged particles, suggesting of a direct correlation between surface charge and protein binding [33, 34]. A study investigating the effect of surface charge density (i.e. zeta potential) of negatively charged polymeric nanoparticles (keeping their size and surface hydrophobicity approximately constant), showed an increase in plasma protein absorption with surface charge density, but showed virtually no difference in the profile of detected protein species [35]. In another report, studies on polystyrene nanoparticles have shown that positively charged particles (bearing basic functional groups) preferentially adsorbed proteins with isoelectric points less than 5.5 (such as albumin), while negatively charged particles (particles with surfaces bearing acidic functional groups) predominantly bound proteins with isoelectric points greater than 5.5 (such as IgG) [36].

The effect of particle size, surface morphology and particle surface area in protein binding has also been investigated. It has been shown that for 50:50, NIPAM/BAM copolymer particles varying in diameter from 70 to 700 nm, the amount of bound protein varied with size and surface morphology. However, the protein pattern remained same for all sizes investigated [37]. Although, particle composition (base material type, shape, and size) influences protein binding; the surface properties (charge and hydrophobicity) are likely to be more important. Binding of proteins to nanoparticles seem to depend on surface chemistry (charge and hydrophobicity); however, the majority of nanoparticles which bind proteins tend to bind the same types of proteins irrespective of their composition (only the amounts of bound protein change).

The hydrophobic nature of nanoparticles surface has also been shown to influence both the amounts of protein bound to the particle, and the type of the bound proteins [37–39]. Usually, hydrophobic particles are opsonized quickly than hydrophilic particles, due to the enhanced absorbability of plasma proteins onto the surface of hydrophobic particles [40, 41]. A less hydrophobic 85:15, NIPAM/BAM copolymer particle was compared to its more hydrophobic counterpart, a 50:50 copolymer particle. The less hydrophobic copolymer particle virtually bound no proteins, except for small amounts of HSA, while the more hydrophobic copolymer particles preferentially bound apolipoproteins AI, AII, AIV, and E as well as HSA, fibrinogen, and various other proteins [37]. With respect to surface chemistry, a study comparing hydrophobic and hydrophilic silica spheres showed a greater change in the secondary structure of fibrinogen and BSA for the hydrophobic particles [42, 43].

3.3.3   Influence of Protein Binding on Biodistribution of Nanoparticles

The size of nanoparticles and the surface morphology of the blood vessel endothelium affect the transport of nanoparticles into surrounding tissues. The endothelia of lung and muscle capillaries are generally characterized by a continuous morphology that allows only small molecules (<3 nm in size) to be transported across the capillary wall, whereas the kidneys have blood vessels with fenestrated endothelia; the liver and spleen have vessels with discontinuous endothelia [44]. Both fenestrated and discontinuous endothelia are associated with larger pore spaces. These pores make it more accessible for nanoparticles with a size of less than 60 nm to enter tissues supplied by such blood vessels [31, 45].

Binding of proteins can lead to change in nanoparticle size, surface morphology and surface charge [46–49]. Such alterations further affect the internalization process of these nanoparticles into macrophages and the overall distribution throughout the body. Binding of certain proteins allow macrophages of the RES to easily recognize nanoparticles [33]. Further, binding of opsonins such as IgG, complement factors, and fibrinogen are correlated with promotion of phagocytosis and subsequent removal of the particles from systemic circulation via cells of the RES. These particles tend to sequester in the RES organs very rapidly and accumulate in the liver and spleen. On the contrary, dysopsonins such as albumin are associated with the promotion of prolonged circulation times in the blood [50–52].

Coating of nanoparticles with specific polymers has been widely investigated to inhibit protein binding, or at least mask the amount, and/or change the overall profile of bound proteins. The incorporation of a Pluronic F127 coating to both single-walled carbon nanotubes and amorphous silica particles enhanced dispersion of the nanoparticles, together with reduction of adsorption of serum proteins, and resulted in reduced toxicity to RAW 264.7 cells [46]. In another study, the introduction of a poloxamine 908 coating to polystyrene nanospheres reduced fibronectin adsorption considerably when compared with the uncoated nanospheres, and also decreased liver accumulation [46, 51]. The most common strategy to prevent nanoparticles from immune recognition is incorporation of PEG, or "PEGylation," [33, 53, 54]. Conjugation of PEG, which is a neutral and amphiphilic molecule, has been shown to mask the interactions of various nanoparticles with blood proteins and prevent recognition by the RES, thereby enhancing the blood circulation time. PEGylation can be achieved by grafting PEG onto the polymer backbone, during block copolymerization, entrapment, or adsorption of PEG chains onto the surface of the particle [53, 55–59]. A study showing comparison of PHDCA and PEG-PHDCA nanoparticles reported that almost twice as much protein was adsorbed onto the non-PEGylated nanoparticle [60]. The composition of adsorbed proteins was also shown to be slightly different.

3.4   Conclusion

Targeted drug delivery systems aim to deliver drugs not only to a specific cell population but also to a specific intracellular compartment. Therefore, it is of immense importance to decipher the mechanism of intracellular trafficking of synthetic nanoparticles that can serve as carriers for drugs or act as therapeutic or imaging agents. Moreover, emphasis should be given to the development of nanoparticles directed towards selected intracellular compartments through engagement of specific cellular trafficking machinery. The published database suggests that the movement of nanoparticles in cells depends on structure and physicochemical characteristics of nanoparticles (size, shape, charge, hydrophobicity, etc.), biospecific interactions between biological moieties attached onto nanoparticles with cells, as well as cell specific endocytosis pathways. The complex mechanism of nanoparticles-cell interactions results in intracellular sorting of nanoparticles toward different destinations and can mediate activation of cellular signaling. Due to the diversity of nanoparticles and cells employed in the trafficking studies, it has become a herculean task to come up with common factors dictating the transport of nanoparticles. The strategy of coating nanoparticles to prevent protein binding and avoid uptake into macrophages, rendering longer circulation times in the body, can be very important for developing a nanoparticle-based therapy limited by nonspecific uptake. On the contrary, specific bound proteins can also help target or direct the nanoparticle towards a particular pathway or area of the body [24]. Henceforth, engineering of nanoparticles that can specifically bind certain proteins of interest for targeting purposes and escape binding of opsonins, can significantly improve the capability to develop targeted nanoparticles for therapeutics.

References

[1] Nimesh, S., M.M. Thibault, M. Lavertu, and M.D. Buschmann. Enhanced gene delivery mediated by low molecular weight chitosan/DNA complexes: effect of pH and serum. *Molecular Biotechnology*, 46: 182–196, 2010.

[2] Schipper, N.G., S. Olsson, J.A. Hoogstraate, A.G. deBoer, K.M. Varum, and P. Artursson. Chitosans as absorption enhancers for poorly absorbable drugs 2: Mechanism of absorption enhancement. *Pharmaceutical Research*, 14: 923–929, 1997.

[3] Thibault, M., S. Nimesh, M. Lavertu, and M.D. Buschmann. Intracellular trafficking and decondensation kinetics of chitosan-pDNA polyplexes. *Molecular Therapy*, 18: 1787–1795, 2010.

[4] Harush-Frenkel, O., E. Rozentur, S. Benita, and Y. Altschuler. Surface charge of nanoparticles determines their endocytic and transcytotic pathway in polarized MDCK cells. *Biomacromolecules*, 9: 435–443, 2008.

[5] Harush-Frenkel, O., N. Debotton, S. Benita, and Y. Altschuler. Targeting of nanoparticles to the clathrin-mediated endocytic pathway. *Biochemical and Biophysical Research Communications*, 353: 26–32, 2007.

[6] Rejman, J., A. Bragonzi, and M. Conese. Role of clathrin- and caveolae-mediated endocytosis in gene transfer mediated by lipo- and polyplexes. *Molecular Therapeutics*, 12: 468–474, 2005.

[7] Sahay, G., J.O. Kim, A.V. Kabanov, and T.K. Bronich. The exploitation of differential endocytic pathways in normal and tumor cells in the selective targeting of nanoparticulate chemotherapeutic agents. *Biomaterials*, 31: 923–933, 2010.

[8] Zhang, L.W. and N.A. Monteiro-Riviere. Mechanisms of quantum dot nanoparticle cellular uptake. *Toxicological Sciences*, 110: 138–155, 2009.

[9] Gratton, S.E., P.A. Ropp, P.D. Pohlhaus, J.C. Luft, V.J. Madden, M.E. Napier, and J.M. DeSimone. The effect of particle design on cellular internalization pathways. *Proceedings of the National Academy of Sciences of the United States of America*, 105: 11613–11618, 2008.

[10] Lai, S.K., K. Hida, C. Chen, and J. Hanes. Characterization of the intracellular dynamics of a nondegradative pathway accessed by polymer nanoparticles. *Journal of Controlled Release*, 125: 107–111, 2008.

[11] Huang, M., Z. Ma, E. Khor, and L.Y. Lim. Uptake of FITC-chitosan nanoparticles by A549 cells. *Pharmaceutical Research*, 19: 1488–1494, 2002.

[12] Champion, J.A., Y.K. Katare, and S. Mitragotri. Making polymeric micro- and nanoparticles of complex shapes. *Proceedings of the National Academy of Sciences of the United States of America*, 104: 11901–11904, 2007.

[13] Champion, J.A. and S. Mitragotri. Role of target geometry in phagocytosis. *Proceedings of the National Academy of Sciences of the United States of America*, 103: 4930–4934, 2006.

[14] Nimesh, S., A. Goyal, V. Pawar, S. Jayaraman, P. Kumar, R. Chandra, Y. Singh, and K.C. Gupta. Polyethylenimine nanoparticles as efficient transfecting agents for mammalian cells. *Journal of Controlled Release*, 110: 457–468, 2006.

[15] Flahaut, E., M.C. Durrieu, M. Remy-Zolghadri, R. Bareille, and C. Baquey. Investigation of the cytotoxicity of CCVD carbon nanotubes towards human umbilical vein endothelial cells. *Carbon*, 44: 1093–1099, 2006.

[16] Monteiro-Riviereand, N.A. and A.O. Inman. Challenges for assessing carbon nanomaterial toxicity to the skin. *Carbon*, 44: 1070–1078, 2006.

[17] Altman, S.A., L. Randers, and G. Rao. Comparison of trypan blue dye exclusion and fluorometric assays for mammalian cell viability determinations. *Biotechnology Progress*, 9: 671–674, 1993.

[18] Bottini, M., S. Bruckner, K. Nika, N. Bottini, S. Bellucci, A. Magrini, A. Bergamaschi, and T. Mustelin. Multi-walled carbon nanotubes induce T lymphocyte apoptosis. *Toxicology Letters*, 160: 121–126, 2006.

[19] Goodman, C.M., C.D. McCusker, T. Yilmaz, and V.M. Rotello. Toxicity of gold nanoparticles functionalized with cationic and anionic side chains. *Bioconjugate Chemistry*, 15: 897–900, 2004.

[20] Mosmann, T. Rapid colorimetric assay for cellular growth and survival: application to proliferation and cytotoxicity assays. *Journal of Immunological Methods*, 65: 55–63, 1983.

[21] Nimesh, S., A. Aggarwal, P. Kumar, Y. Singh, K.C. Gupta, and R. Chandra. Influence of acyl chain length on transfection mediated by acylated PEI nanoparticles. *International Journal of Pharmaceutics*, 337: 265–274, 2007.

[22] Nimesh, S. and R. Chandra. Polyethylenimine nanoparticles as an efficient in vitro siRNA delivery system. *European Journal of Pharmaceutics and Biopharmaceutics*, 73: 43–49, 2009.

[23] Nimesh, S. and R. Chandra. Guanidinium-grafted polyethylenimine: an efficient trans-fecting agent for mammalian cells. *European Journal of Pharmaceutics and Biopharma-ceutics*, 68: 647–655, 2008.

[24] Patnaik, S., A. Aggarwal, S. Nimesh, A. Goel, M. Ganguli, N. Saini, Y. Singh, and K.C. Gupta. PEI-alginate nanocomposites as efficient in vitro gene transfection agents. *Journal of Controlled Release*, 114: 398–409, 2006.

[25] Nociari, M.M., A. Shalev, P. Benias, and C. Russo. A novel one-step, highly sensitive fluo-rometric assay to evaluate cell-mediated cytotoxicity. *Journal of Immunological Methods*, 213: 157–167, 1998.

[26] Nakayama, G.R., M.C. Caton, M.P. Nova, and Z. Parandoosh. Assessment of the alamar blue assay for cellular growth and viability in vitro. *Journal of Immunological Methods*, 204: 205–208, 1997.

[27] Agasti, S.S., S. Rana, M.H. Park, C.K. Kim, C.C. You, and V.M. Rotello. Nanoparticles for detection and diagnosis. *Advanced Drug Delivery Reviews*, 62: 316–328, 2010.

[28] Kennedy, L.C., L.R. Bickford, N.A. Lewinski, A.J. Coughlin, Y. Hu, E.S. Day, J.L. West, and R.A. Drezek. A new era for cancer treatment: Gold-nanoparticle-mediated thermal therapies. *Small*, 7: 169–183, 2011.

[29] Veiseh, O., J.W. Gunn, and M. Zhang. Design and fabrication of magnetic nanoparticles for targeted drug delivery and imaging. *Advanced Drug Delivery Reviews*, 62: 284–304, 2010.

[30] Zolnik, B.S. and N. Sadrieh. Regulatory perspective on the importance of ADME assess-ment of nanoscale material containing drugs. *Advanced Drug Delivery Reviews*, 61: 422–427, 2009.

[31] Li, M., K.T. Al-Jamal, K. Kostarelos, and J. Reineke. Physiologically based pharmacoki-netic modeling of nanoparticles. *ACS Nano*, 4: 6303–6317, 2010.

[32] Alexis, F., E. Pridgen, L.K. Molnar, and O.C. Farokhzad. Factors affecting the clearance and biodistribution of polymeric nanoparticles. *Molecular Pharmaceutics*, 5: 505–515, 2008.

[33] Owens, D.E., 3rd and N.A. Peppas. Opsonization, biodistribution, and pharmacokinetics of polymeric nanoparticles. *International Journal of Pharmaceutics*, 307: 93–102, 2006.

[34] Roser, M., D. Fischer, and T. Kissel. Surface-modified biodegradable albumin nano- and microspheres. II: Effect of surface charges on in vitro phagocytosis and biodistribution in rats. *European Journal of Pharmaceutics and Biopharmaceutics*, 46: 255–263, 1998.

[35] Gessner, A., A. Lieske, B. Paulke, and R. Muller. Influence of surface charge density on protein adsorption on polymeric nanoparticles: Analysis by two-dimensional elec-trophoresis. *European Journal of Pharmaceutics and Biopharmaceutics*, 54: 165–170, 2002.

[36] Gessner, A., A. Lieske, B.R. Paulke and R.H. Muller. Functional groups on polystyrene model nanoparticles: Influence on protein adsorption. *Journal of Biomedial Materials Research Part A*, 65: 319–326, 2003.

[37] Cedervall, T., I. Lynch, M. Foy, T. Berggard, S.C. Donnelly, G. Cagney, S. Linse, and K.A. Dawson. Detailed identification of plasma proteins adsorbed on copolymer nanoparticles. *Angewandte Chemie International Edition in English*, 46: 5754–5756, 2007.

[38] Luck, M., B.R. Paulke, W. Schroder, T. Blunk, and R.H. Muller. Analysis of plasma pro-tein adsorption on polymeric nanoparticles with different surface characteristics. *Journal of Biomedial Materials Research*, 39: 478–485, 1998.

[39] Goppert T.M. and R.H. Muller. Plasma protein adsorption of tween 80- and poloxamer 188-stabilized solid lipid nanoparticles. *Journal of Drug Targeting*, 11: 225–231, 2003.

[40] Carrstensen, H., R.H. Muller, and B.W. Muller. Particle size, surface hydrophobicity and interaction with serum of parenteral fat emulsions and model drug carriers as parameters related to RES uptake. *Clinical Nutrition*, 11: 289–297, 1992.

[41] Norman, M.E., P. Williams, and L. Illum. Human serum albumin as a probe for surface conditioning (opsonization) of block copolymer-coated microspheres. *Biomaterials*, 13: 841–849, 1992.

[42] Asuri, P., S.S. Bale, S.S. Karajanagi, and R.S. Kane. The protein-nanomaterial interface. *Current Opinion in Biotechnology*, 17: 562–568, 2006.

[43] Roach, P., D. Farrar, and C.C. Perry. Surface tailoring for controlled protein adsorption: Effect of topography at the nanometer scale and chemistry. *Journal of the American Chemical Society*, 128: 3939–3945, 2006.

[44] Chrastina, A., K.A. Massey, and J.E. Schnitzer. Overcoming in vivo barriers to targeted nanodelivery. *Wiley Interdisciplinary Reviews: Nanomedicine and Nanobiotechnology*, 3: 421–437, 2011.

[45] Choi, H.S. and J.V. Frangioni. Nanoparticles for biomedical imaging: fundamentals of clinical translation. *Molecular Imaging*, 9: 291–310, 2010.

[46] Dutta, D., S.K. Sundaram, J.G. Teeguarden, B.J. Riley, L.S. Fifield, J.M. Jacobs, S.R. Addleman, G.A. Kaysen, B.M. Moudgil, and T.J. Weber. Adsorbed proteins influence the biological activity and molecular targeting of nanomaterials. *Toxicological Sciences*, 100: 303–315, 2007.

[47] Chithrani, B.D., A.A. Ghazani, and W.C. Chan. Determining the size and shape dependence of gold nanoparticle uptake into mammalian cells. *Nano Letters*, 6: 662–668, 2006.

[48] Moghimi, S.M., A.C. Hunter, and J.C. Murray. Long-circulating and target-specific nanoparticles: theory to practice. *Pharmacol Rev.*, 53: 283–318, 2001.

[49] Nagayama, S., K. Ogawara, Y. Fukuoka, K. Higaki, and T. Kimura. Time-dependent changes in opsonin amount associated on nanoparticles alter their hepatic uptake characteristics. *International Journal of Pharmaceutics*, 342: 215–221, 2007.

[50] Goppert, T.M. and R.H. Muller. Adsorption kinetics of plasma proteins on solid lipid nanoparticles for drug targeting. *International Journal of Pharmaceutics*, 302: 172–186, 2005.

[51] Moghimi, S.M., I.S. Muir, L. Illum, S.S. Davis, and V. Kolb-Bachofen. Coating particles with a block co-polymer (poloxamine-908) suppresses opsonization but permits the activity of dysopsonins in the serum. *Biochimica et Biophysica Acta*, 1179: 157–165, 1993.

[52] Ogawara, K., K. Furumoto, S. Nagayama, K. Minato, K. Higaki, T. Kai, and T. Kimura. Pre-coating with serum albumin reduces receptor-mediated hepatic disposition of polystyrene nanosphere: Implications for rational design of nanoparticles. *Journal of Controlled Release*, 100: 451–455, 2004.

[53] Gref, R., M. Luck, P. Quellec, M. Marchand, E. Dellacherie, S. Harnisch, T. Blunk, and R.H. Muller. 'Stealth' corona-core nanoparticles surface modified by polyethylene glycol (PEG): influences of the corona (PEG chain length and surface density) and of the core composition on phagocytic uptake and plasma protein adsorption. *Colloids and Surfaces B:* Biointerfaces, 18: 301–313, 2000.

[54] Peracchia, M.T., S. Harnisch, H. Pinto-Alphandary, A. Gulik, J.C. Dedieu, D. Desmaele, J. d'Angelo, R.H. Muller, and P. Couvreur. Visualization of in vitro protein-rejecting properties of PEGylated stealth polycyanoacrylate nanoparticles. *Biomaterials*, 20: 1269–1275, 1999.

[55] Brus, C., H. Petersen, A. Aigner, F. Czubayko, and T. Kissel. Physicochemical and biological characterization of polyethylenimine-graft-poly(ethylene glycol) block copolymers as a delivery system for oligonucleotides and ribozymes. *Bioconjugate Chemistry*, 15: 677–684, 2004.

[56] Brus, C., H. Petersen, A. Aigner, F. Czubayko, and T. Kissel. Efficiency of polyethylenimines and polyethylenimine-graft-poly (ethylene glycol) block copolymers to protect oligonucleotides against enzymatic degradation. *European Journal of Pharmaceutics and Biopharmaceutics*, 57: 427–430, 2004.

[57] Chen, J., B. Tian, X. Yin, Y. Zhang, D. Hu, Z. Hu, M. Liu, Y. Pan, J. Zhao, H. Li, C. Hou, and J. Wang. Preparation, characterization and transfection efficiency of cationic PEGylated PLA nanoparticles as gene delivery systems. *Journal of Biotechnology*, 130: 107–113, 2007.

[58] Cheng, J., B.A. Teply, I. Sherifi, J. Sung, G. Luther, F.X. Gu, E. Levy-Nissenbaum, A.F. Radovic-Moreno, R. Langer, and O.C. Farokhzad. Formulation of functionalized PLGA-PEG nanoparticles for in vivo targeted drug delivery. *Biomaterials*, 28: 869–876, 2007.

[59] Choi, Y.H., F. Liu, J.S. Kim, Y.K. Choi, J.S. Park, and S.W. Kim. Polyethylene glycol-grafted poly-L-lysine as polymeric gene carrier. *Journal of Controlled Release*, 54: 39–48, 1998.

[60] Kim, H.R., K. Andrieux, C. Delomenie, H. Chacun, M. Appel, D. Desmaele, F. Taran, D. Georgin, P. Couvreur, and M. Taverna. Analysis of plasma protein adsorption onto PEGylated nanoparticles by complementary methods: 2-DE, CE and Protein Lab-on-chip system. *Electrophoresis*, 28: 2252–2261, 2007.

4

Concept and Barriers to Gene Silencing

4.1 Introduction

RNA interference (RNAi), initially discovered by Fire *et al.* in the nematode *Caenorhabditis elegans*, is a mechanism that inhibits gene expression at the stage of translation or by hindering the transcription of specific genes [1]. Gene silencing mediated by RNAi has been suggested as an approach for efficient knockdown of gene expression in eukaryotic organisms, and accelerated the discovery of a unifying mechanism that underlies a host of cellular and developmental pathways and therefore, defines gene function in mammalian cells. It is a highly conserved mechanism that controls the selective posttranscriptional down-regulation of target genes in the cells. RNAi is mediated by small sequences of 21 to 23 nucleotide double stranded RNAs, referred to as siRNAs. These siRNAs have the potential to degrade messenger RNAs (mRNAs) that are complementary to one of the siRNA strands. RNAi has found numerous applications not only in biological research and drug development but also *in vivo* inhibition of gene expression related to human diseases. Cancer is a severe disease characterized by uncontrolled growth of cells, invasion and metastasis. The most common treatment options for cancer include surgery, chemotherapy and radiotherapy. Recently, RNAi technology has been tested as an efficient therapy for the down-regulation of defective genes [2]. siRNA bears several advantages over conventional chemotherapy such as reduced non-specific tissues toxicity and high target specificity. However, the therapeutic application of siRNA is hindered by poor cellular uptake and rapid nuclease degradation [3]. Hence, there is a

S. Nimesh and R. Chandra,
Theory, Techniques and Applications of Nanotechnology in Gene Silencing, 47–60.
© 2011 *River Publishers. All rights reserved.*

need of a delivery vehicle capable of administering siRNA efficiently, safely, and repeatedly to *in vitro* and *in vivo* milieu. The carrier system should provide protection against degradative nucleases, possess target specificity, capable of intracellular uptake and escape from the endosome/lysosome into cytosol leading to efficient gene knockdown. Although, viral vectors have been explored by several groups for siRNA delivery, safety concerns such as chances of oncogenicity, inflammation and immunogenicity hampers their clinical application [4, 5]. Nanomedicine-based vectors have been widely investigated as potential candidates for effective siRNA delivery.

Nanomedicine is a wide arena that collectively holds polymeric micelles, quantum dots, liposomes, polymer-drug conjugates, dendrimers, biodegradable nanoparticles, inorganic nanoparticles and other materials in nanometer size range with therapeutic relevance. One of the most promising candidates of nanomedicine is nanoparticles that have emerged with numerous applications in the field of targeted drug and gene delivery. The small size of nanoparticles has contributed to their better tissue penetration and targeting [6]. Nanoparticles prepared from polycationic polymers have been largely explored to deliver DNA and siRNA. Several natural and synthetic polycationic polymers including chitosan, PEI have been utilized for preparing nanoparticles to deliver siRNA.

4.2   Basic Principle and Mechanism of RNAi

The mRNA degradation by RNAi is probably the most powerful and specific mechanism for gene silencing. Double-stranded RNA (dsRNA) is an important regulator of gene expression which triggers different types of gene silencing that are collectively referred to as RNA interference. The mechanism of RNAi takes place in the cytoplasm of the cells (Figure 4.1). It is a multistep process which initiates with the binding of dsRNA-specific endonuclease (a cytoplasmic ribonuclease III (RNase III)-like protein) called Dicer that is capable of cleaving long dsRNAs [7]. This binding of dicer is followed by cleavage of long dsRNA into smaller duplexes with 19 paired nucleotides and two nucleotide overhangs at both 3' ends and these small dsRNAs hence generated are called siRNAs [8]. These siRNA molecules further bind to a nuclease containing multiprotein complex called RNA-induced silencing complex (RISC) [9]. The RISC complex by virtue of its RNA helicase activity unbinds the double

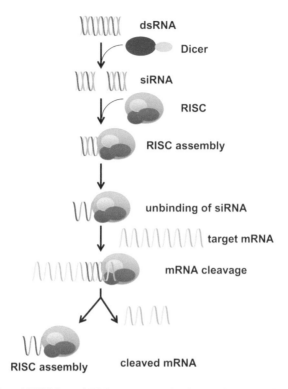

Fig. 4.1 Mechanism of RNAi. Long dsRNA precursor molecules are cleaved by Dicer yielding siRNA. siRNA incorporates into RISC assembly, followed by unwinding of the double-stranded molecule by the helicase activity of RISC. The sense strand of siRNA is removed and the antisense strand binds targeted mRNA, which is then cleaved by RISC and subsequently degraded by cellular nucleases.

stranded siRNA molecule, where the sense strand is removed and is further degraded by cellular nucleases [10]. The antisense strand of siRNA remains bound to the RISC complex and is directed to the target mRNA sequence, where it anneals complementarily by Watson-Crick base pairing. Finally, the RISC complex cleaves the target mRNA by endonucleolytic activity thereby preventing its translation into protein [11]. At a later stage, the cleaved products are released and degraded by cellular nucleases, leaving the unengaged RISC complex to further search for additional target mRNAs and carry out its activity of interference [12]. A more complete understanding of the process is imminent as understanding the mechanisms of RNA silencing may shed more light on various diseases. Tuschl *et al.* were first to report that siRNAs possess the capability of silencing targeted gene in mammalian cells. The study evidenced that siRNA mediates sequence specific gene silencing and the dicing step can be further bypassed by transfection of siRNA molecules into cells [8]. RNAi has been found to be more efficacious than other nucleic acids as it utilizes the cellular machinery for gene silencing. Furthermore, siRNAs are more stable than antisense oligonucleotides against nuclease degradation and hence have prolonged therapeutic effects as compared to antisense therapy [13, 14].

4.3 Barriers to siRNA Delivery

The therapeutic success of siRNA-based gene silencing relies on efficient delivery to the target sites, which in turn depends on the delivery vectors. However, during the delivery process, siRNA faces numerous barriers that hamper their efficacy (Figure 4.2). These barriers can be divided into two categories, as discussed in the following sections:

4.3.1 *In Vitro* Barriers

Although siRNA functions in cytoplasm while DNA is expressed in nucleus, the delivery of siRNA faces almost the same obstacles as DNA delivery from cell targeting to internalization and endosomal escape [15, 16]. Complexation of siRNA with polycationic polymers results in condensation of siRNA due to electrostatic interaction between the polymer amino groups and siRNA phosphate groups. The size of the complexes thus formed depends on various factors such as concentration of siRNA, pH, type of buffer, and N/P ratio. It has been reported that vector-siRNA complexes having size range of 50–200 nm mostly enter cells by endocytosis or pinocytosis [17, 18]. The aggregation of complexes in the extracellular environment turns out to be one of the first major obstacles to overcome. For systems employing cationic polymers to deliver siRNA, aggregation of the complexes is frequently observed with complexes prepared near charge neutrality. In the presence of excess positive charge, they also often manifest reduced colloidal stability. The increased ionic strength typically encountered on introduction of these systems into the biological milieu (relative to formulation values) may also have significant effects on their physical properties. To avoid aggregation and instability, the surface of nanoparticles is manipulated by incorporating PEG or sugar molecules (e.g. cyclodextrin and hyaluronic acid) [19, 20]. Efficient uptake of siRNA by cells is hampered due to the presence of polyanionic charge

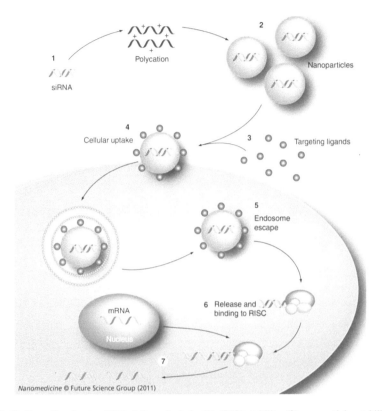

Fig. 4.2 Outline of barriers to siRNA delivery include: (1) siRNA stability, (2) nanoparticles stability, (3) nanoparticles targeting, (4) nanoparticles internalization, (5) endosomal escape of nanoparticles, (6) release of siRNA from nanoparticles and (7) siRNA off-target effects. Adapted from Nimesh *et al.* [54].

and its large size (~13 kDa). The initial modifications to enhance uptake of siRNA employed cell penetrating peptides (CPPs) and cholesterol [21, 22]. Later, various polycationic formulations comprising PEI, polyarginine, PLL, histidylated PLL, chitosan and poly(amidoamine) PAMAM dendrimers have been used to deliver siRNA.

Internalization of complexes is followed by fusion of intracellular vesicles carrying these complexes with the endocytic vesicles. These early endosomes fuse with sorting endosomes, which in turn transfer their contents to the late endosomes. Late endosomal vesicles are acidified (pH 5–6) by membrane-bound proton pump ATPases. The endosomal content is then relocated to the lysosomes, which are further acidified (pH ~4.5) and contain various nucleases that promote the degradation of the siRNAs. To avoid lysosomal degradation, siRNAs (free or complexed with the carrier) must escape from the endosome into the cytosol, where they can associate with the RNAi machinery. Endosomal escape is a major barrier for efficient siRNA delivery. Several hypotheses have been proposed for the endosomal escape of complexes. One common mechanism suggests physical disruption of the negatively charged endosomal membrane by direct interaction with the cationic polymer e.g. PAMAM dendrimers and PLL [23]. Other strategies exploit either pH-responsive or reduction-sensitive polymers. PEI consists of ionizable amino groups and favors the release of endocytosed polyplexes by proton sponge effect [24–27]. Kim *et al.* reported that poly (PAsp(DET)) undergoes efficient endosomal escape due to pH-dependent protonation of N-(2-aminoethyl)-2-aminoethyl groups in the PAsp(DET) side chain [28]. Block copolymer of dimethylaminoethyl methacrylate (DMAEMA) and propylacrylic acid (PAA) which undergoes structural rearrangements in response to endosomal pH have also been investigated for siRNA delivery [29]. The block consists of a cationic complexation component and PAA that mediates a hydrophilic-to-hydrophobic transition at the endosomal pH leading to membrane disruption. Thermosensitive cationic polymeric nanocapsules based on a temperature induced swelling ($\sim$119 nm at 37°C and $\sim$ 412 at 15°C), which results in physical disruption of the endosome have also been employed to deliver siRNA [30]. As an alternative strategy, short peptides that are capable of translocating throughout biological membranes called CPPs have also been used to modify polyplexes to improve endosomal release of siRNAs [31]. Photochemical internalization (PCI) technology has also been used to enhance the release of endocytosed macromolecules to the cytosol [32]. This technology is based on the activation of endocytosed photosensitive molecules (photosensitizers) by light to induce the release of endocytic vesicle contents before they are transferred to the lysosome. PCI has been employed to facilitate endosomal escape of siRNA targeting epidermal growth factor receptor (EGFR) [33].

Stability of nanoparticles is desirable for extracellular siRNA protection; however, decomplexation is essential for siRNA to mediate gene silencing. Hence, a subtle balance between the protection and release of siRNA is required for efficient silencing with nanoparticles [34, 35]. Bioresponsive nanoparticles where the release of siRNA occurs in response to intracellular stimulus have been used to facilitate this process. Incorporation of acid-labile ketal linkages to PEI enhances transfection and RNAi. Ketalized PEI-siRNA polyplexes showed much higher silencing efficiency than unmodified linear-PEI via selective cytoplasmic localization of the polyplexes and efficient disassembly of siRNA from the polyplexes [36]. Polyplexes containing reducible polycations that degrade in response to cytoplasmic redox conditions have also been investigated for siRNA delivery. Reducible poly(amidoethylenimine) efficiently condenses siRNA to form stable complexes under physiological conditions and results in complete release of siRNA in a reductive environment [37]. PIC micelle prepared with a disulfide crosslinked core through the assembly of iminothiolane-modified PEG-*block*-PLL [PEG-*b*-(PLL-IM)] and siRNA showed 100-fold higher siRNA transfection efficacy compared with non-crosslinked PICs [38]. This higher efficiency is due to selective release of siRNA in intracellular milieu while protecting siRNA in extracellular milieu from degradation and non-specific clearance by PICs. Also, siRNA-grafted to poly(aspartic acid) [PAsp(-SS-siRNA)] *via* a disulfide linkage gave higher siRNA efficiency due to efficient siRNA release from the PIC under intracellular reductive conditions [39].

4.3.2 *In Vivo* barriers

The extracellular barriers to siRNA depend on the route of administration e.g. intravenous, intranasal, intratracheal, subcutaneous, intratumor, intramuscular, or oral, which in turn, depends upon the targeted disease. One of the most significant biological barriers encountered by systemically administered siRNA is the nuclease activity in plasma. The major enzymatic activity that occurs in the plasma is the 3′ exonuclease; although, cleavage of internucleotide bonds can also take place. Chemical modification can significantly improve the stability of oligonucleotides in the biological milieu, as well as allowing improvements in selectivity and reduced toxicity. Several chemical modifications have been proposed to protect siRNA in serum, which includes modification of the sugars or backbone of siRNA by 2′-O-methyl and 2′-deoxy-2′-fluoro (OMe/F) or phosphorothioate linkages [40]. However, siRNA complexed with polymeric nanoparticles easily bypass this barrier as it is no more available for 3′ nuclease binding followed by cleavage.

The pharmacokinetics and biodistribution of siRNA has also been extensively investigated. Several animal studies report that the biodistribution of siRNA duplexes is similar to that of single stranded antisense molecules, with highest uptake in kidney followed by liver [41–43]. However, the high renal uptake has been associated with strong effects of the siRNA in "knocking down" target molecules in that tissue [43]. Various studies have observed that siRNA biodistribution in animals is done by employing radioactive tagging, or RNase protection assays, or simple HPLC [42, 44, 45]. Incorporation of cholesterol, tocopherol, or other lipid moieties improve binding of the oligonucleotide to serum lipoproteins and/or albumin [44, 46, 47]. This results in enhanced circulation lifetimes and, more importantly, in enhanced hepatic uptake *via* the low density lipoprotein receptor.

Rapid clearance by RES is another major problem against efficient siRNA delivery. Phagocytic cells of RES, more specifically the Kupffer cells in the liver and splenic macrophages, can endocytose siRNA oligonucleotides, as well as carriers used to deliver them [48]. Nanoparticles used to deliver siRNA are considered as foreign particles in the body. Hence, they are recognized by opsonins which are composed of immunoglobulins, and complement system proteins, and other serum proteins. These opsonized particles are recognized by a variety of receptors present on the cell surface of macrophages. Immunoglobulin G-opsonized particles are recognized by Fc receptors and complement-opsonized particles are internalized through complement receptors thereby leading to their degradation. However, modulation of nanoparticle surfaces with hydrophilic polymers such as PEG, reduces the adsorption of opsonins followed by reduced clearance by phagocytosis [49]. One of the most crucial steps involved with siRNA is the specific recognition and binding to the target mRNA in the cytosol, to mediate silencing. Some of the off-target effects observed include toxicity, stimulation of immune response, inflammation, undesirable effects on other genes [50]. Chemical modification of siRNA where uridine bases are replaced by their $2'$-fluoro, $2'$-deoxy, or $2'-O$-methyl modified counterparts have been reported to block immune recognition of siRNAs by Toll Like Receptors (TLR) [51]. Polymers used to deliver siRNA, which avoid the delivery into the endosomes or block TLR signalling, have been proposed as another strategy.

The endothelial cells lining the vascular lumen pose both a barrier and an opportunity for oligonucleotide-based therapeutics. The large size and negative charge of siRNA molecules makes it difficult to pass across capillary endothelial cells by simple diffusion. However, the fenestrated capillary endothelium of the liver, and the glomerular pores within the kidney may well allow siRNA to pass into these tissues. For other organs, the continuous capillary wall may be appearing as an important barrier to silencing. Henceforth, this barrier is likely to be a major contributor to the failure of these agents after intravenous administration. Hydrodynamic injection increases the diameter of hepatic fenestrations, improves extravasation and thus favors hepatic gene silencing [52]. PEG conjugated to PEI nanoparticles have been shown to significantly improve the movement of siRNA into the lung, liver and tumors [53]. Tumor vasculature possess unusual properties which can be exploited by macromolecules > 40 kDa (e.g. complexed siRNA) and be retained within the tumor. In principle, various factors should be considered both at the cellular and whole organism level to design efficient siRNA delivery vectors.

4.4 Conclusions

Although many issues of stability and efficacy have largely been overcome, yet efficient intracellular delivery remains a key problem in the way of effective oligonucleotide therapeutics. For both oligonucleotides and those conjugated to nanoparticles, entrapment in endosomes and inefficient release to the cytosol remain an unresolved issue. Although modifications such as PEGylation has improved the circulation time of nanoparticles, the RES still manages to capture a significant amount of the injected dose. Stability and cellular uptake may need to be optimal during trafficking for siRNA to meet its potential as a therapeutic modality.

References

[1] Fire, A., S.Q. Xu, M.K. Montgomery, S.A. Kostas, S.E. Driver, and C.C. Mello. Potent and specific genetic interference by double-stranded RNA in caenorhabditis elegans. *Nature*, 391: 806–811, 1998.

[2] Huh, M.S., S.Y. Lee, S. Park, S. Lee, H. Chung, Y. Choi, Y.K. Oh, J.H. Park, S.Y. Jeong, K. Choi, K. Kim, and I.C. Kwon. Tumor-homing glycol chitosan/polyethylenimine nanoparticles for the systemic delivery of siRNA in tumor-bearing mice. *Journal of Controlled Release*, 144: 134–143, 2010.

[3] Zhang, S., B. Zhao, H. Jiang, B. Wang, and B. Ma. Cationic lipids and polymers mediated vectors for delivery of siRNA. *Journal of Controlled Release*, 123: 1–10, 2007.

[4] Xu, D., D. McCarty, A. Fernandes, M. Fisher, R.J. Samulski, and R.L. Juliano. Delivery of MDR1 small interfering RNA by self-complementary recombinant adeno-associated virus vector. *Molecular Therapeutics*, 11: 523–530, 2005.

[5] Zaiss, A.K. and D.A. Muruve. Immune responses to adeno-associated virus vectors. *Current Gene Therapy*, 5: 323–331, 2005.

[6] Peer, D., J.M. Karp, S. Hong, O.C. Farokhzad, R. Margalit, and R. Langer. Nanocarriers as an emerging platform for cancer therapy. *Nature Nanotechnology*, 2: 751–760, 2007.

[7] E. Bernstein, A.A. Caudy, S.M. Hammond, and G.J. Hannon. Role for a bidentate ribonuclease in the initiation step of RNA interference. *Nature*, 409: 363–366, 2001.

[8] Elbashir, S.M., J. Harborth, W. Lendeckel, A. Yalcin, K. Weber, and T. Tuschl. Duplexes of 21-nucleotide RNAs mediate RNA interference in cultured mammalian cells. *Nature*, 411: 494–498, 2001.

[9] Hammond, S.M., E. Bernstein, D. Beach, and G.J. Hannon. An RNA-directed nuclease mediates post-transcriptional gene silencing in drosophila cells. *Nature* 404: 293–296, 2000.

[10] Nykanen, A., B. Haley, and P.D. Zamore. ATP requirements and small interfering RNA structure in the RNA interference pathway. *Cell*, 107: 309–321, 2001.

[11] Martinez, J. and T. Tuschl. RISC is a 5' phosphomonoester-producing RNA endonuclease. *Genes and Development*, 18: 975–980, 2004.

[12] Leung, R.K.M. and P.A. Whittaker. RNA interference: From gene silencing to gene-specific therapeutics. *Pharmacology & Therapeutics*, 107: 222–239, 2005.

[13] Jana, S., C. Chakraborty, S. Nandi, and J.K. Deb. RNA interference: Potential therapeutic targets. *Applied Microbiology and Biotechnology*, 65: 649–657, 2004.

[14] Li, C.X., A. Parker, E. Menocal, S. Xiang, L. Borodyansky, and J.H. Fruehauf. Delivery of RNA interference. *Cell Cycle*, 5: 2103–2109, 2006.

[15] Grayson, A.C., A.M. Doody, and D. Putnam. Biophysical and structural characterization of polyethylenimine-mediated siRNA delivery *in vitro*. *Pharmaceutical Research*, 23: 1868–1876, 2006.

[16] Katas, H. and H.O. Alpar. Development and characterisation of chitosan nanoparticles for siRNA delivery. *Journal of Controlled Release*, 115: 216–225, 2006.

[17] Gao, H., W. Shi, and L.B. Freund. Mechanics of receptor-mediated endocytosis. *Proceedings of the National Academy of Sciences of the United States of America*, 102: 9469–9474, 2005.

[18] Aoki, H., M. Satoh, K. Mitsuzuka, A. Ito, S. Saito, T. Funato, M. Endoh, T. Takahashi, and Y. Arai. Inhibition of motility and invasiveness of renal cell carcinoma induced by short interfering RNA transfection of beta 1,4GalNAc transferase. *FEBS Letters*, 567: 203–208, 2004.

[19] Bartlett, D.W., H. Su, I.J. Hildebrandt, W.A. Weber, and M.E. Davis. Impact of tumor-specific targeting on the biodistribution and efficacy of siRNA nanoparticles measured by multimodality *in vivo* imaging. *Proceedings of the National Academy of Sciences of the United States of America*, 104: 15549–15554, 2007.

[20] Lee, H., H. Mok, S. Lee, Y.K. Oh, and T.G. Park. Target-specific intracellular delivery of siRNA using degradable hyaluronic acid nanogels. *Journal of Controlled Release*, 119: 245–252, 2007.

[21] Moschos, S.A., S.W. Jones, M.M. Perry, A.E. Williams, J.S. Erjefalt, J.J. Turner, P.J. Barnes, B.S. Sproat, M.J. Gait, and M.A. Lindsay. Lung delivery studies using

siRNA conjugated to TAT(48-60) and penetratin reveal peptide induced reduction in gene expression and induction of innate immunity. *Bioconjugate chemistry*, 18: 1450–1459, 2007.

[22] Muratovska, A. and M.R. Eccles. Conjugate for efficient delivery of short interfering RNA (siRNA) into mammalian cells. *FEBS Letters*, 558: 63–68, 2004.

[23] Zhang, Z.Y. and B.D. Smith. High-generation polycationic dendrimers are unusually effective at disrupting anionic vesicles: Membrane bending model. *Bioconjugate chemistry*, 11: 805–814, 2000.

[24] Boussif, O., F. Lezoualc'h, M.A. Zanta, M.D. Mergny, D. Scherman, B. Demeneix, and J.P. Behr. A versatile vector for gene and oligonucleotide transfer into cells in culture and in vivo: Polyethylenimine. *Proceedings of the National Academy of Sciences of the United States of America*, 92: 7297–7301, 1995.

[25] Thomas, M., J.J. Lu, Q. Ge, C. Zhang, J. Chen, and A.M. Klibanov. Full deacylation of polyethylenimine dramatically boosts its gene delivery efficiency and specificity to mouse lung. *Proceedings of the National Academy of Sciences of the United States of America*, 102: 5679–5684, 2005.

[26] Thomas, M. and A.M. Klibanov. Enhancing polyethylenimine's delivery of plasmid DNA into mammalian cells. *Proceedings of the National Academy of Sciences of the United States of America*, 99: 14640–14645, 2002.

[27] Akinc, A., M. Thomas, A.M. Klibanov, and R. Langer. Exploring polyethylenimine-mediated DNA transfection and the proton sponge hypothesis. *Journal of Gene Medicine*, 7: 657–663, 2005.

[28] Kim, H.J., A. Ishii, K. Miyata, Y. Lee, S. Wu, M. Oba, N. Nishiyama, and K. Kataoka. Introduction of stearoyl moieties into a biocompatible cationic polyaspartamide derivative, PAsp(DET), with endosomal escaping function for enhanced siRNA-mediated gene knockdown. *Journal of Controlled Release*, 145: 141–148, 2010.

[29] Convertine, A.J., D.S. Benoit, C.L. Duvall, A.S. Hoffman, and P.S. Stayton. Development of a novel endosomolytic diblock copolymer for siRNA delivery. *Journal of Controlled Release*, 133: 221–229, 2009.

[30] Lee, S.H., S.H. Choi, S.H. Kim, and T.G. Park. Thermally sensitive cationic polymer nanocapsules for specific cytosolic delivery and efficient gene silencing of siRNA: Swelling induced physical disruption of endosome by cold shock. *Journal of Controlled Release*, 125: 25–32, 2008.

[31] Endoh, T. and T. Ohtsuki. Cellular siRNA delivery using cell-penetrating peptides modified for endosomal escape. *Advanced Drug Delivery Reviews*, 61: 704–709, 2009.

[32] Berg, K., P.K. Selbo, L. Prasmickaite, T.E. Tjelle, K. Sandvig, J. Moan, G. Gaudernack, O. Fodstad, S. Kjolsrud, H. Anholt, G.H. Rodal, S.K. Rodal, and A. Hogset. Photochemical internalization: A novel technology for delivery of macromolecules into cytosol. *Cancer Research*, 59: 1180–1183, 1999.

[33] Oliveira, S., I. van Rooy, O. Kranenburg, G. Storm, and R.M. Schiffelers. Fusogenic peptides enhance endosomal escape improving siRNA-induced silencing of oncogenes. *International Journal of Pharmaceutics* 331: 211–214, 2007.

[34] Zhang, K., H. Fang, Z. Wang, J.S. Taylor, and K.L. Wooley. Cationic shell-crosslinked knedel-like nanoparticles for highly efficient gene and oligonucleotide transfection of mammalian cells. *Biomaterials*, 30: 968–977, 2009.

[35] Zhang, K., H. Fang, Z. Wang, Z. Li, J.S. Taylor, and K.L. Wooley. Structure-activity relationships of cationic shell-crosslinked knedel-like nanoparticles: Shell composition and transfection efficiency/cytotoxicity. *Biomaterials*, 31: 1805–1813, 2010.

[36] Shim, M.S. and Y.J. Kwon. Acid-responsive linear polyethylenimine for efficient, specific, and biocompatible siRNA delivery. *Bioconjugate Chemistry*, 20: 488–499, 2009.

[37] Hoon Jeong, J., L.V. Christensen, J.W. Yockman, Z. Zhong, J.F. Engbersen, W. Jong Kim, J. Feijen, and S. Wan Kim. Reducible poly(amido ethylenimine) directed to enhance RNA interference. *Biomaterials*, 28: 1912–1917, 2007.

[38] Matsumoto, S., R.J. Christie, N. Nishiyama, K. Miyata, A. Ishii, M. Oba, H. Koyama, Y. Yamasaki, and K. Kataoka. Environment-responsive block copolymer micelles with a disulfide cross-linked core for enhanced siRNA delivery. *Biomacromolecules*, 10: 119–127, 2009.

[39] Takemoto, H., A. Ishii, K. Miyata, M. Nakanishi, M. Oba, T. Ishii, Y. Yamasaki, N. Nishiyama and K. Kataoka. Polyion complex stability and gene silencing efficiency with a siRNA-grafted polymer delivery system. *Biomaterials*, 31: 8097–8105, 2010.

[40] Akhtar, S. and I.F. Benter. Nonviral delivery of synthetic siRNAs *in vivo*. *Journal of Clinical Investigation*, 117: 3623–3632, 2007.

[41] Merkel, O.M., D. Librizzi, A. Pfestroff, T. Schurrat, M. Behe and T. Kissel. In vivo SPECT and real-time gamma camera imaging of biodistribution and pharmacokinetics of siRNA delivery using an optimized radiolabeling and purification procedure. *Bioconjugate Chemistry*, 20: 174–182, 2009.

[42] Braasch, D.A., Z. Paroo, A. Constantinescu, G. Ren, O.K. Oz, R.P. Mason, and D.R. Corey. Biodistribution of phosphodiester and phosphorothioate siRNA. *Bioorganic and Medicinal Chemistry Letters*, 14: 1139–1143, 2004.

[43] van de Water, F.M., O.C. Boerman, A.C. Wouterse, J.G. Peters, F.G. Russel, and R. Masereeuw. Intravenously administered short interfering RNA accumulates in the kidney and selectively suppresses gene function in renal proximal tubules. *Drug Metabolism and Disposition*, 34: 1393–1397, 2006.

[44] Soutschek, J., A. Akinc, B. Bramlage, K. Charisse, R. Constien, M. Donoghue, S. Elbashir, A. Geick, P. Hadwiger, J. Harborth, M. John, V. Kesavan, G. Lavine, R.K. Pandey, T. Racie, K.G. Rajeev, I. Rohl, I. Toudjarska, G. Wang, S. Wuschko, D. Bumcrot, V. Koteliansky, S. Limmer, M. Manoharan, and H.P. Vornlocher. Therapeutic silencing of an endogenous gene by systemic administration of modified siRNAs. *Nature*, 432: 173–178, 2004.

[45] Zimmermann, T.S., A.C. Lee, A. Akinc, B. Bramlage, D. Bumcrot, M.N. Fedoruk, J. Harborth, J.A. Heyes, L.B. Jeffs, M. John, A.D. Judge, K. Lam, K. McClintock, L.V. Nechev, L.R. Palmer, T. Racie, I. Rohl, S. Seiffert, S. Shanmugam, V. Sood, J. Soutschek, I. Toudjarska, A.J. Wheat, E. Yaworski, W. Zedalis, V. Koteliansky, M. Manoharan, H.P. Vornlocher, and I. MacLachlan. RNAi-mediated gene silencing in non-human primates. *Nature*, 441: 111–114, 2006.

[46] Nishina, K., T. Unno, Y. Uno, T. Kubodera, T. Kanouchi, H. Mizusawa, and T. Yokota. Efficient in vivo delivery of siRNA to the liver by conjugation of alpha-tocopherol. *Molecular Therapy*, 16: 734–740, 2008.

[47] Wolfrum, C., S. Shi, K.N. Jayaprakash, M. Jayaraman, G. Wang, R.K. Pandey, K.G. Rajeev, T. Nakayama, K. Charrise, E.M. Ndungo, T. Zimmermann, V. Koteliansky,

M. Manoharan, and M. Stoffel. Mechanisms and optimization of *in vivo* delivery of lipophilic siRNAs. *Nature Biotechnology*, 25: 1149–1157, 2007.

[48] Alexis, F., E. Pridgen, L.K. Molnar, and O.C. Farokhzad. Factors affecting the clearance and biodistribution of polymeric nanoparticles. *Molecular Pharmaceutics*, 5: 505–515, 2008.

[49] Van Vlerken, L.E., T.K. Vyas, and M.M. Amiji. Poly(ethylene glycol)-modified nanocarriers for tumor-targeted and intracellular delivery. *Pharmaceutical Research*, 24: 1405–1414, 2007.

[50] Aigner, A. Gene silencing through RNA interference (RNAi) in vivo: Strategies based on the direct application of siRNAs. *Journal of Biotechnology*, 124: 12–25, 2006.

[51] Sioud, M. Single-stranded small interfering RNA are more immunostimulatory than their double-stranded counterparts: A central role for 2'-hydroxyl uridines in immune responses. *European Journal of Immunology*, 36: 1222–1230, 2006.

[52] Zhang, G., X. Gao, Y.K. Song, R. Vollmer, D.B. Stolz, J.Z. Gasiorowski, D.A. Dean, and D. Liu. Hydroporation as the mechanism of hydrodynamic delivery. *Gene Therapy*, 11: 675–682, 2004.

[53] Schiffelers, R.M., A. Ansari, J. Xu, Q. Zhou, Q. Tang, G. Storm, G. Molema, P.Y. Lu, P.V. Scaria, and M.C. Woodle. Cancer siRNA therapy by tumor selective delivery with ligand-targeted sterically stabilized nanoparticle. *Nucleic Acids Research*, 32:e149 2004.

[54] Nimesh, S., N. Gupta, and R. Chandra. Strategies and advances in nanomedicine for targeted siRNA delivery. *Nanomedicine*, 6: 729–746, 2011.

5

Nanoparticles-Mediated Targeted siRNA Delivery

5.1 Introduction

One of the most challenging tasks in the development of therapeutic siRNAs is the lack of specificity of siRNA for tissues or cells. In clinical practice, the potential approach for successful applications of siRNA is to deliver at the target cells and tissues. The targeted delivery not only minimizes the chances of adverse effects, but also reduces the amount of doses required to achieve the desired therapeutic effect. Cell-specific ligands that include antibodies (Abs), sugar molecules, vitamins, and hormones have been largely employed to confer cell specificity on siRNA delivery systems. Cellular uptake of ligands is generally enhanced by the aid of receptor-mediated endocytosis if the specific interaction between the ligand and its membranous receptor occurs. Therefore, with a much lower dose, an enhanced therapeutic efficacy can be achieved with increase in the specific cellular uptake by targeted delivery.

Strategies to deploy siRNA that target tumor cells appear to hold promise with focus on the potential therapeutic benefits for the treatment as well as to reduce the adverse side effects associated with cancer therapy. In many cases, cancer cells have unique antigen(s) and/or receptor(s) on the cell surface that are not found on the normal cells. Targeted therapeutics require the construction of molecules that specifically recognize these macromolecules and deliver therapeutic siRNA to the target cells. Nanocarriers could be functionalized with biomolecules for "active" tumor targeting (Figure 5.1). Surface ligands include antibodies, aptamers, peptides, or small molecules which

S. Nimesh and R. Chandra,
Theory, Techniques and Applications of Nanotechnology in Gene Silencing, 61–70.

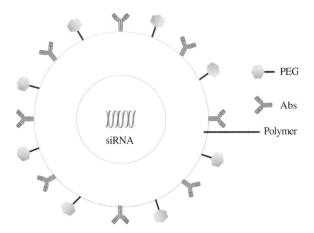

Fig. 5.1   Schematic strategies for targeted siRNA delivery. The siRNA is condensed within cationic polymers such as PEI, chitosan etc. that form nanoparticles, and the surface of the nanoparticles is tagged with PEG and targeting moieties such as Abs.

recognize tumor-specific or tumor-associated antigens in the tumor microenvironment [1–6]. The active targeting mechanism takes advantage of highly specific interactions between the targeting ligand and certain tissues or cell surface antigens to increase cellular uptake and increase tumor retention. Conjugation approaches have been developed to control the amount of targeting ligands on the surface of the nanoparticles. In the case of weak binding ligands, multivalent functionalization on the surface of the nanoparticles provides sufficient avidity. In general, small molecule ligands such as peptides, sugars, and small molecules are more attractive than antibodies due to higher stability, higher purity, ease of production through synthetic routes, and non-immunogenicity.

Two common approaches have been greatly exploited for receptor-mediated targeting. The first strategy is to target the tumor microenvironment that include the extracellular matrix or surface receptors on endothelial cells of the tumor blood vessels, which is usually most efficient for the delivery of immune induction or antiangiogenesis molecules. The second approach is to target surface receptors of the tumor cells for intracellular delivery of cytotoxic agents or signal-pathway inhibitors. Nanocarriers targeted towards the extracellular markers of transmembrane tumor antigens are generally taken up by cancer cells via receptor-mediated endocytosis, leading to efficient intracellular delivery of therapeutic loads. Although it is difficult to compare the efficacies of two approaches, in recent times, receptor targeted nanocarriers have shown promising results.

5.2 Strategies for Targeted siRNA Delivery

Tissues or cells can be specifically targeted using cell specific affinity moieties such as peptides or antibodies. Targeted delivery of siRNA to the desired site not only increases its efficacy but also prevents possible off-target side effects. The major step in targeting is that the cellular receptors should be efficiently internalized after binding to their ligands, followed by rapid reappearance on the cell surface to allow repeated targeting. Another key parameter is to design ligands for efficient siRNA delivery to the cell surface receptors. A number of proteins are expressed at the cell surface levels by the cells during various tumor stages, which can be exploited to modulate delivery system for target specificity. The various strategies for targeting siRNA delivery are discussed below in this section.

5.2.1 Antibodies

Antibody-mediated targeted siRNA delivery has attracted much attention due to its high *in vivo* stability and specificity to the surface protein of target cell. In the last several decades mAbs have been widely investigated as targeting molecules. Engineered mAbs have been commonly used for siRNA delivery purposes to reach the targeted gene. For efficient functional activity of the engineered Abs in the human body, they have to escape the immune system. Recent development of mAbs has thus been centered on chimeric, partially humanized and fully humanized derivatives to minimize their immunogenicity. Various engineered mAbs have been successfully employed to target disease, for example; rituximab, trastuzumab, cetuximab, and bevacizumab.

Song *et al.* designed a protamine-antibody fusion protein to deliver siRNA to HIV-infected or envelope transfected cells [7]. The fusion protein (F105-P) was designed with the protamine coding sequence linked to the C terminus of the heavy chain Fab fragment of an HIV-1 envelope antibody. siR-NAs bound to F105-P induced silencing in cells that expressed the HIV-1 envelope. Additionally, siRNAs targeted against the HIV-1 capsid gene gag, inhibited HIV replication in hard-to-transfect, HIV-infected primary T cells.

Intratumoral or intravenous injection of F105-P-complexed siRNAs into mice targeted the HIV envelope-expressing B16 melanoma cells, but not the normal tissue or envelope-negative B16 cells; injection of F105-P with siRNAs targeting the c-myc, murine double minute 2 (MDM2) and vascular endothelial growth factor (VEGF) inhibited the envelope-expressing subcutaneous B16 tumors. Furthermore, an ErbB2 single-chain antibody fused with protamine delivered siRNAs specifically into ErbB2-expressing cancer cells. Chen *et al.* developed a liposome-polycation-hyaluronic acid (LPH) nanoparticle formulation modified with tumor targeting single chain variable antibody fragment (scFv) for systemic delivery of siRNA and microRNA into experimental lung metastasis of murine B16F10 melanoma. The siRNAs delivered by the scFv-tagged nanoparticles efficiently downregulated the target genes (c-Myc/MDM2/VEGF) in the lung metastasis [8]. mAbs generated against oncogenes such as HER2 or VEGF have been widely used for cancer treatments. Recently, Gao *et al.* synthesized PEGylated DOPE immunoliposomes conjugated with the Fab' of recombinant humanized anti-HER2 mAbs (PIL) for siRNA delivery [9]. The results demonstrated that the lyophilized PIL prepared by the lyophilization/rehydration, containing 2.5% PEG showed the best HER1 gene silencing activity.

Despite showing very promising results in the targeted disease therapeutics, there are quite a few limitations associated with mAbs usage. They are large, complex macromolecules that require significant engineering at the molecular level to be effective. The production cost is also very high and there is variation from batch to batch, which limits their efficiency as tagging molecules. Despite these limitations and challenges with usage of mAbs, there is an increased interest in using Abs fragments as tagging molecules while retaining the high antigen binding specificity of antibodies. These moieties include the Fab fragments, scFv, minibodies, diabodies, and nanobodies. The Fab fragment consists of a constant and a variable domain of each of the heavy and the light chains; scFv is a fusion of the variable regions of the heavy and light chains. Minibodies are engineered antibody fragment that is a fusion between scFV and a C_H3 domain that self-assembles into a bivalent dimer. Diabodies are covalently linked dimers or noncovalent dimers of scFvs. Nanobodies, which are the smallest of all fully functional antigen-binding fragments, evolved from the variable domain of heavy-chain antibodies. Nanobodies are derived from single-domain antibodies, carrying only a functional heavy chain without the light chain. These antibody fragments are engineered to retain high affinity for target antigens but possess less immunogenicity and a smaller size, and thus are better suited for molecular targeting.

5.2.2 Peptides

Peptides are attractive targeting molecules due to their smaller size, lower immunogenicity, higher stability and ease of manufacture. The development of peptide phage libraries ($\sim 10^{11}$ different peptide sequences), bacterial peptide display libraries, plasmid peptide libraries, and new screening technologies have made their selection much easier, contributing to their popularity as targeting ligands. Small peptides such as a RGD peptide targeting the integrin expression, which is upregulated at sites of neovasculature, have been employed due to their specificity [10]. Self-assembling nanoparticles of PEGylated PEI with an RGD peptide ligand attached at the distal end of the PEG were prepared with siRNA against the vascular endothelial growth factor receptor-2 (VEGFR2) [10]. Cell delivery and activity of PEGylated PEI was found to be siRNA sequence-specific and dependent on the presence of peptide ligand. Intravenous administration into tumor-bearing mice gave selective uptake, siRNA sequence-specific inhibition of protein expression within the tumor, and inhibition of both tumor angiogenesis and its growth rate. In another study, nanosized particles were formed upon complexation of siRNA with the cationic polymer RGD-PEG-PEI [11]. Both the circulation kinetics and the overall tumor localization of the siRNA complexes were similar to noncomplexed siRNA. However, the carrier changed the intratumoral distribution of siRNA within the tumor. Waite *et al.* prepared generation 5, PAMAM dendrimers modified by the addition of cyclic RGD targeting peptides and evaluated their ability to associate with siRNA and mediate siRNA delivery to U87 malignant glioma cells [12]. PAMAM-RGD conjugates were able to complex with siRNA to form complexes of approximately 200 nm in size. Modest siRNA delivery was observed in U87 cells using either PAMAM or PAMAM-RGD conjugates. PAMAM-RGD conjugates prevented the adhesion of U87 cells to fibrinogen-coated plates, in a manner that depends on the number of RGD ligands per dendrimer. The delivery of siRNA through three-dimensional multicellular spheroids of U87 cells was enhanced using PAMAM-RGD conjugates compared to the native PAMAM dendrimers, presumably by interfering with integrin-ECM contacts present in a three-dimensional tumor model. RGD peptide conjugated to chitosan nanoparticles (RGD-CH-NP) by thiolation reaction was used to target $\alpha v \beta 3$ integrin [13]. RGD-CH-NP loaded with siRNA significantly increased selective intratumoral delivery in orthotopic animal models of ovarian cancer. These nanoparticles were employed to deliver PLXDC1-targeted siRNA into the $\alpha v \beta 3$ integrin positive tumor endothelial cells in the A2780 tumor bearing mice. Targeted carrier was developed by the conjugation of the $\alpha v \beta 3 / \alpha v \beta 5$ integrin-binding RGD peptide (ACDCRGDCFC) to the cationic polymer, branched PEI, with a hydrophilic PEG spacer [14]. PEI-g-PEG-RGD was used to deliver siRNA against VEGFR1 into tumor site. The physicochemical properties of PEI-g-PEG-RGD/siRNA complexes were evaluated. Further, tumor growth profile investigation after systemic administration of PEI-g-PEG-RGD/siRNA complexes suggest of significant silencing of VEGR1 gene.

5.2.3   Transferrin

The distribution of TFR is known to increase on the malignant cell surfaces in many types of tumors thereby providing a cell surface receptor that can be targeted. The intracellular movement of TFR-mediated endocytosis has been well investigated. The endocytosis involves the recycling of TFR to the cell surface without its intracellular degradation. As compared to antibodies, transferrin used for targeting TFR is more affordable and is commercially available in large quantities. Hence, transferrin-mediated delivery has been extensively explored for a variety of targets, including tumors, endothelial cells and the brain [48]. The studies involve the chemical coupling of either transferrin itself or of antiTFR antibodies or antibody fragments to the surface of nanoparticles. siRNA condensed with cationic cyclodextrin polymers to form nanoparticles followed by conjugation of transferrin was used for targeting Ewing's sarcoma tumor cells that express high levels of the TFR [15, 16]. Sequence-specific knockdown of EWS-FLI 1 and the corresponding inhibition of tumor growth were observed in mice after tail vein injection of the siRNA at a 2.5 mg/kg dose. Interestingly, prolonged systemic administration of the CDP delivery system did not induce detectable immune response. Moreover, transferrin-targeted nanoparticles were found to be safe in nonhuman primates for M2 siRNA delivery [49]. A recent study reported nanoparticles containing a linear CDP, transferrin, PEG and siRNA to knockdown the expression of the RRM2 [17]. Tumor biopsies from melanoma treated patients showed the presence of intracellularly localized nanoparticles and reduction in both the mRNA and the protein levels of RRM2. Also, in A/J mice bearing subcutaneous Neuro2A tumours treated with intravenous injection of siRNA against RRM2 containing transferrin conjugated CDP nanoparticles showed decrease in tumor growth [18].

5.2.4   Other Targeting Ligands

Small molecules such as folate, carbohydrates and cholesterol have been reported as targeting moieties in several studies. FRs are over-expressed on the cell surfaces in many different types of cancers. To enhance the intracellular delivery of shRNA and reduce cytotoxicity, low MW PEI derivatized to folate-chitosan-graft-PEI (FC-g-PEI) copolymer [19]. Aerosol delivery of FC-g-PEI/Akt1 shRNA complexes suppressed lung tumorigenesis in a urethane induced lung cancer mouse model through the Akt signaling pathway. Kim *et al.* delivered antisense oligodeoxynucleotide, synthetic siRNA and plasmid siRNA to inhibit GFP using PEI-PEG-folate conjugates in FRs over-expressing KB cells [20]. Among these three nucleic acids, synthetic siRNA exhibited the most dose-effective and fastest gene silencing effect.

The asialoglycoprotein receptor (ASGP-R) recognizes many different types of carbohydrates and is highly expressed in liver cells. Nie *et al.* employed galactosylated PEG-graft-PEI (Gal-PEG-PEI) to deliver psiRNA to knockdown HLA-E gene expression in HepG2 cells. Real time PCR analysis and western blot analysis revealed upto 60% inhibition of HLA-E gene expression, 48 h post-transfection [21]. Cholesterol and its analogues have also been exploited as targeting ligands as these are recognized by cholesterol receptors on hepatocytes. Watanabe *et al.* reported that intravenous delivery of apolipoprotein (Apo)B- specific siRNA with a sixth generation of dendritic PLL resulted in siRNA-mediated knockdown of ApoB in healthy C57BL/6 mice without hepatotoxicity and observed significant reduction in serum low density lipoprotein cholesterol in ApoE deficient mice [22].

5.3   Conclusions

The development of ligand derivatized targeted nanoparticles as therapeutic agents have generated great enthusiasm in both academia and industry. Targeted nanoparticles used for siRNA delivery have shown promising results in preclinical studies, demonstrating their potential as therapeutics carriers. However, several challenges remain in to be answered. The nanoparticles with intracellular uptake, higher target tissue concentration, improved efficacy and lower toxicity compared with non-targeted nanoparticles, will be the ultimate choice as delivery vehicles for therapeutic agents. Further optimization of the nanoparticles surface and size, as well as the targeting ligand, will pave the way towards development of better targeted nanoparticle systems.

References

[1] Alexis, F., E. Pridgen, L.K. Molnar, and O.C. Farokhzad. Factors affecting the clearance and biodistribution of polymeric nanoparticles. *Molecular. Pharmaceutics*, 5: 505–515, 2008.

[2] Bareford, L.M. and P.W. Swaan. Endocytic mechanisms for targeted drug delivery. *Advanced Drug Delivery Reviews*, 59: 748–758, 2007.

[3] Farokhzad, O.C., J. Cheng, B.A. Teply, I. Sherifi, S. Jon, P.W. Kantoff, J.P. Richie, and R. Langer. Targeted nanoparticle-aptamer bioconjugates for cancer chemotherapy in vivo. *Proceedings of the National Academy of Sciences of the United States of America*, 103: 6315–6320, 2006.

[4] Farokhzad, O.C., J.M. Karp, and R. Langer. Nanoparticle-aptamer bioconjugates for cancer targeting. *Expert Opinion on Drug Delivery*, 3: 311–324, 2006.

[5] Sudimack, J. and R.J. Lee. Targeted drug delivery via the folate receptor. *Advanced Drug Delivery Reviews*, 41: 147–162, 2000.

[6] Van Vlerken, L.E. and M.M. Amiji. Multi-functional polymeric nanoparticles for tumour-targeted drug delivery. *Expert Opinion on Drug Delivery*, 3: 205–216, 2006.

[7] Song, E., P. Zhu, S.K. Lee, D. Chowdhury, S. Kussman, D.M. Dykxhoorn, Y. Feng, D. Palliser, D.B. Weiner, P. Shankar, W.A. Marasco, and J. Lieberman. Antibody mediated in vivo delivery of small interfering RNAs via cell-surface receptors. *Nature Biotechnology*, 23: 709–717, 2005.

[8] Chen, Y., X. Zhu, X. Zhang, B. Liu, and L. Huang. Nanoparticles modified with tumor-targeting scFv deliver siRNA and miRNA for cancer therapy. *Molecular Therapy*, 18: 1650–1656, 2010.

[9] Gao, J., J. Sun, H. Li, W. Liu, Y. Zhang, B. Li, W. Qian, H. Wang, J. Chen, and Y. Guo. Lyophilized HER2-specific PEGylated immunoliposomes for active siRNA gene silencing. *Biomaterials*, 31: 2655–2664, 2010.

[10] Schiffelers, R.M., A. Ansari, J. Xu, Q. Zhou, Q. Tang, G. Storm, G. Molema, P.Y. Lu, P.V. Scaria, and M.C. Woodle. Cancer siRNA therapy by tumor selective delivery with ligand-targeted sterically stabilized nanoparticle. *Nucleic Acids Research*, 32: e149, 2004.

[11] de Wolf, H.K., C.J. Snel, F.J. Verbaan, R.M. Schiffelers, W.E. Hennink, and G. Storm. Effect of cationic carriers on the pharmacokinetics and tumor localization of nucleic acids after intravenous administration. *International Journal of Pharmaceutics*, 331: 167–175, 2007.

[12] Waite, C.L. and C.M. Roth. PAMAM-RGD conjugates enhance siRNA delivery through a multicellular spheroid model of malignant glioma. *Bioconjugate Chemistry*, 20: 1908–1916, 2009.

[13] Han, H.D., L.S. Mangala, J.W. Lee, M.M. Shahzad, H.S. Kim, D. Shen, E.J. Nam, E.M. Mora, R.L. Stone, C. Lu, S.J. Lee, J.W. Roh, A.M. Nick, G. Lopez-Berestein, and A.K. Sood. Targeted gene silencing using RGD-labeled chitosan nanoparticles. *Clinical Cancer Research*, 16: 3910–3922, 2010.

[14] Kim, J., S.W. Kim, and W.J. Kim. PEI-g-PEG-RGD/small interference RNA polyplex-mediated silencing of vascular endothelial growth factor receptor and its potential as an anti-angiogenic tumor therapeutic strategy. *Oligonucleotides*, 21: 101–107, 2011.

[15] Heidel, J.D., Z. Yu, J.Y. Liu, S.M. Rele, Y. Liang, R.K. Zeidan, D.J. Kornbrust, and M.E. Davis. Administration in nonhuman primates of escalating intravenous doses of targeted nanoparticles containing ribonucleotide reductase subunit M2 siRNA. *Proceedings of the National Academy of Sciences of the United States of America*, 104: 5715–5721, 2007.

[16] Hu-Lieskovan, S., J.D. Heidel, D.W. Bartlett, M.E. Davis, and T.J. Triche. Sequence-specific knockdown of EWS-FLI1 by targeted, nonviral delivery of small interfering RNA inhibits tumor growth in a murine model of metastatic ewing's sarcoma. *Cancer Research*, 65: 8984–8992, 2005.

[17] Davis, M.E., J.E. Zuckerman, C.H.J. Choi, D. Seligson, A. Tolcher, C.A. Alabi, Y. Yen, J.D. Heidel, and A. Ribas. Evidence of RNAi in humans from systemically administered siRNA via targeted nanoparticles. *Nature*, 464: 1067–1070, 2010.

[18] Bartlett, D.W., H. Su, I.J. Hildebrandt, W.A. Weber, and M.E. Davis. Impact of tumor-specific targeting on the biodistribution and efficacy of siRNA nanoparticles measured by multimodality *in vivo* imaging. *Proceedings of the National Academy of Sciences of the United States of America*, 104: 15549–15554, 2007.

[19] Jiang, H.L., C.X. Xu, Y.K. Kim, R. Arote, D. Jere, H.T. Lim, M.H. Cho, and C.S. Cho. The suppression of lung tumorigenesis by aerosol-delivered folate-chitosan-graft-polyethylenimine/Akt1 shRNA complexes through the akt signaling pathway. *Biomaterials*, 30: 5844–5852, 2009.

[20] Kim, S.H., H. Mok, J.H. Jeong, S.W. Kim, and T.G. Park. Comparative evaluation of target-specific GFP gene silencing efficiencies for antisense ODN, synthetic siRNA, and siRNA plasmid complexed with PEI-PEG-FOL conjugate. *Bioconjugate Chemistry*, 17: 241–244, 2006.

[21] Nie, C., C. Liu, G. Chen, J. Dai, H. Li, and X. Shuai. Hepatocyte-targeted psiRNA delivery mediated by galactosylated poly(ethylene glycol)-graft-polyethylenimine in vitro. *Journal of Biomaterials Applications*, 2010.

[22] Watanabe, K., M. Harada-Shiba, A. Suzuki, R. Gokuden, R. Kurihara, Y. Sugao, T. Mori, Y. Katayama, and T. Niidome. *In vivo* siRNA delivery with dendritic poly(l-lysine) for the treatment of hypercholesterolemia. *Molecular BioSystems*, 5: 1306–1310, 2009.

6

Polymers for siRNA Delivery

6.1 Introduction

As the DNA and siRNA have similar physicochemical properties, vectors developed for DNA have also been employed to deliver siRNA. Both the linear as well as branched cationic polymers are efficient DNA transfection agents. The structural and chemical properties of these polymers have been investigated in detail in this chapter. The positively charged polymers form polyplexes via electrostatic interactions with the negatively charged phosphates of DNA [1]. This interaction results in DNA condensation and protection of plasmids from nuclease digestion. Similarly siRNA polymer polyplexes or nanoparticles mediate siRNA delivery (Figure 6.1). Other proposed polymeric vectors of siRNA include micelles, nanoplexes, nanocapsules, and nanogels [2]. The properties of polyplexes (e.g. size, surface charge, and structure) are dependent on the ratio of the positive charges of cationic polymers to the number of phosphate groups of siRNA (i.e. N/P ratio). Several natural and synthetic polymers such as PEI, PLL, PLGA, poly(alkylcyanoacrylate), chitosan, and gelatin have been investigated. In this book, we have focused on some of the widely exploited cationic polymers for *in vitro* and *in vivo* siRNA delivery.

6.1.1 Advantages and Limitations of Nanoparticles

Nanoparticles, being compact, are well suited to traverse cellular membranes to mediate gene delivery. It is also expected that due to smaller size, nanoparticles would be less susceptible to reticuloendothelial system clearance and

S. Nimesh and R. Chandra,
Theory, Techniques and Applications of Nanotechnology in Gene Silencing, 71–78.
© 2011 *River Publishers. All rights reserved.*

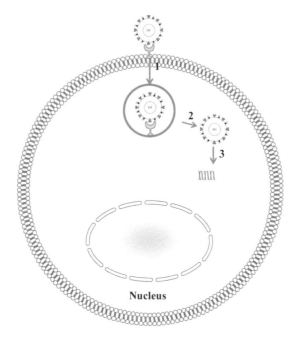

Fig. 6.1   Intracellular pathway of nanoparticle-mediated siRNA delivery. Receptor-mediated endocytosis (1) leading to entrapment into endosomes, followed by release at low pH of the nanoparticles (2) and finally release of siRNA from nanoparticles.

will have better penetration into tissues and cells, when used in *in vivo* therapy. Nanoparticles prepared from polymers bear several advantages such as ease of manipulation with scope to changing the MW, geometry (linear and branched), stability, safety, low cost and high flexibility regarding the size of transgene delivered. Nanoparticles can easily be tagged with various targeting moieties such as RGD peptides or transferrin to achieve target specific siRNA delivery [3, 4].

Further, owing to their size (usually 10 to 200 nm), nanoparticles can readily interact with surface biomolecules of cell or inside the cell. Also, due to their small size, nanoparticles can penetrate tissues such as tumors in depth with a high level of specificity, thereby improving the targeted delivery of drug/gene [5]. Despite being so advantageous and successful implication in numerous studies, polymeric nanoparticles have some limitations. The polycationic polymers constituting nanoparticles undergo strong electrostatic interaction with plasma membrane proteins, which can lead to destabilization and ultimately rupture of the cell membrane [6]. A comparative study between polycationic, neutral and polyanionic polymers revealed that the polycationic polymers have the highest toxicity followed by neutral and anionic ones [7]. Strategies based on reduction of surface charge by coating with hyaluronic acid or PEG have been investigated to circumvent this problem [8, 9]. Among the linear and branched PEIs, the latter are more toxic and less suitable for transfection, particularly at higher N/P ratios [10]. Additionally, PEI is a non-biodegradable polymer that can accumulate within the cells interfering with vital intracellular processes [11, 12]. A complement system can be activated by PEI/DNA complexes if the ratio of cation to anion is high, but the extent of its activation is lowered as the PEI/DNA complexes approach neutrality [13, 14]. Chemical modifications where small MW PEIs are coupled to generate large higher MW molecules using degradable bifunctional linkages have been found useful in reducing toxicity. Chitosan has been found to be an efficient *in vitro* and *in vivo* siRNA delivery vector capable of mediating gene silencing with no toxicity [15, 16]. However, it suffers from low transfection efficiency and optimum conditions for transfection are not clear, and therefore, the results must be well elucidated before clinical applications. Moreover, non-specific stimulation of innate immune inflammatory response such as type I interferon (IFN) production, associated with RNA duplex interaction with endosomal Toll-like receptors (TLR), can be potentiated with nanoparticle-siRNA delivery into the endosomes [17]. However, $2'$-O-methyl nucleotides have been incorporated into siRNA duplex strand for interruption of TLR-7 interaction and associated non-specific effects [18].

High sequence-specificity of RNA interference is dependent on the specific binding of siRNA to its target mRNA such that the nucleolytic activity of the RISC complex is initiated. However, in a real milieu, off-target effects have been indicated where non-specific RNAi-induced gene silencing occurred with the introduction of a gene-specific siRNA. Thus, partial sequence similarity between non-targeted mRNA and siRNA led to cross reactivity. In fact, in some cases, regions comprising of only 11–15 contiguous nucleotides of sequence identity were sufficient to induce gene silencing. However, the prediction of these off-target activities is so far difficult.

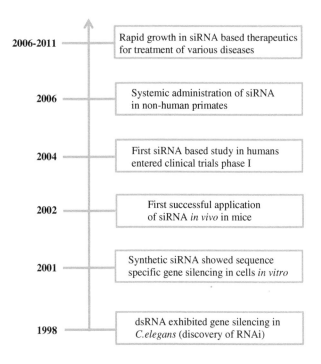

Fig. 6.2 Progressive milestones in the field of RNAi-mediated gene silencing.

6.2   Present Clinical Scenario

The identification of new therapeutic targets and hence the strategy to treat them is more straightforward using siRNA than traditional pharmaceuticals, which led to rapid development of siRNA-based therapeutics. Within 10 years after the discovery of the RNAi mechanism, some of the lead siRNA candidates have reached clinical trials (Figure 6.2) [19]. The first set of clinical trials based on siRNA technology for treatment of wet neovascular age-related macular degeneration (AMD), using siRNA targeting the VEGF signaling pathway, is underway. Phase I and II trials to assess the safety, tolerability, and efficacy of the siRNA C and 5 (bevasiranib; Opko Health Inc.), that target VEGF in AMD patients have been completed and currently, phase III trials are underway. Another siRNA-027 (termed AGN211745; Sirna Therapeutics), a chemically modified siRNA-targeting VEGF receptor 1 has entered phase II trials and promises to provide significant clinical improvement in visual acuity, with no serious adverse events or dose-limiting toxicities, in a relevant subset of patients. Another AMD siRNA candidate, RTP-801i, blocks REDD-1 gene expression and has been investigated in phase I clinical trials. The first siRNA against respiratory viral infection is the intranasal ALN-RSV01 of Alnylam Pharmaceuticals, which targets respiratory syncytial virus (RSV). ALN-RSV01 is currently being investigated in phase II trials in naturally infected adult patients [20]. For treatment of chronic hepatitis B virus (HBV) infection, Nucleonics has begun a phase 1 human safety study of NUC B1000. NUC B1000 is a siRNA- based product designed to reduce the malignant effects of hepatitis and is systemically administered using a cationic lipid formulation.

Two siRNA-based therapeutics, CALAA-01 and Atu027, have entered clinical phase trials for the treatment of solid tumors; both agents are administered by intravenous injection. CALAA-01 is a siRNA targeting the RRM2 where siRNA is formulated in self-assembled cyclodextrin nanoparticles with surface pegylation and in conjugation with the transferrin ligand, meant to target the TRs in tumor cells [4]. The first in-human phase I trials of intravenous injection of CALAA-01 in patients with solid tumors had begun in May 2008. The findings suggested successful delivery of nanoparticles to intracellular localizations and reduction of corresponding mRNA and protein levels in tumor biopsies. This happens to be the first evidence of specific gene inhibition by siRNA in three patients after systemic administration [21]. Atu027 is a siRNA lipoplex designed to target protein kinase N3 [22]. Preclinical studies indicated that repeated intravenous administrations of Atu027 resulted in specific, RNAi-mediated silencing of protein kinase N3 expression in mice, rats, and nonhuman primates, significant inhibition of tumor growth, and inhibition of the formation of lymph node metastasis in orthotopic mouse models of prostate and pancreatic cancers (98). The phase I clinical trials of Atu027 in patients with advanced solid cancer were initiated in July 2009.

6.3 Conclusions

Though *in vivo* siRNA delivery has made significant amount of progress, many challenges are yet to be overcome. There is a need to emphasize on strategies to minimize off-target effects and immune stimulation. Recently, polymer-based synthetic siRNA delivery systems have shown promising results in gene knockdown by RNAi after systemic administration. Target specific RNAi effects have been demonstrated in non-human primates after intravenous administration of siRNA using appropriate delivery systems. However, the delivery of siRNA remains a key issue towards the therapeutic achievement using RNAi. Biologically potent siRNA must be transferred safely and efficiently to malignant target tissues to mediate silencing of target genes by an RNAi process. Hence, effective and optimized delivery systems for therapeutic siRNA will provide indispensible strategy for cancer therapy. Moreover, encouraging clinical results pave the way for transitioning RNAi from a research table into clinic evaluation and future applications.

References

[1] Agarwal, A., R. Unfer, and S.K. Mallapragada. Novel cationic pentablock copolymers as non-viral vectors for gene therapy. *Journal of Controlled Release*, 103: 245–258, 2005.

[2] de Martimprey, H., C. Vauthier, C. Malvy, and P. Couvreur. Polymer nanocarriers for the delivery of small fragments of nucleic acids: Oligonucleotides and siRNA. *European Journal of Pharmaceutics and Biopharmaceutics*, 71: 490–504, 2009.

[3] Schiffelers, R.M., A. Ansari, J. Xu, Q. Zhou, Q. Tang, G. Storm, G. Molema, P.Y. Lu, P.V. Scaria, and M.C. Woodle. Cancer siRNA therapy by tumor selective delivery with ligand-targeted sterically stabilized nanoparticle. *Nucleic Acids Research*, 32: e149, 2004.

[4] Davis, M.E. The first targeted delivery of siRNA in humans via a self-assembling, cyclodextrin polymer-based nanoparticle: From concept to clinic. *Mol Pharm.*, 6: 659–668, 2009.

[5] Cuenca, A.G., H. Jiang, S.N. Hochwald, M. Delano, W.G. Cance, and S.R. Grobmyer. Emerging implications of nanotechnology on cancer diagnostics and therapeutics. *Cancer*, 107: 459–466, 2006.

[6] Fischer, D., Y. Li, B. Ahlemeyer, J. Krieglstein, and T. Kissel. In vitro cytotoxicity testing of polycations: Influence of polymer structure on cell viability and hemolysis. *Biomaterials*, 24: 1121–1131, 2003.

[7] Jevprasesphant, R., J. Penny, R. Jalal, D. Attwood, N.B. McKeown, and A. D'Emanuele. The influence of surface modification on the cytotoxicity of PAMAM dendrimers. *International Journal of Pharmaceutics*, 252: 263–266, 2003.

[8] Jiang, G., K. Park, J. Kim, K.S. Kim, E.J. Oh, H. Kang, S.E. Han, Y.K. Oh, T.G. Park, and S. Kwang Hahn. Hyaluronic acid-polyethyleneimine conjugate for target specific intracellular delivery of siRNA. *Biopolymers*, 89: 635–642, 2008.

[9] Duan, Y., C. Yang, Z. Zhang, J. Liu, J. Zheng, and D. Kong. Poly(ethylene glycol)-grafted polyethylenimine modified with G250 monoclonal antibody for tumor gene therapy. *Human Gene Therapy*, 21: 191–198, 2010.

[10] Wightman, L., R. Kircheis, V. Rossler, S. Carotta, R. Ruzicka, M. Kursa, and E. Wagner. Different behavior of branched and linear polyethylenimine for gene delivery in vitro and in vivo. *Journal of Gene Medicine*, 3: 362–372, 2001.

[11] Godbey, W.T., K.K. Wu, G.J. Hirasaki, and A.G. Mikos. Improved packing of poly(ethylenimine)/DNA complexes increases transfection efficiency. *Gene Therapy*, 6: 1380–1388, 1999.

[12] Godbey, W.T., K.K. Wu, and A.G. Mikos. Poly(ethylenimine)-mediated gene delivery affects endothelial cell function and viability. *Biomaterials*, 22: 471–480, 2001.

[13] Ferrari, S., E. Moro, A. Pettenazzo, J.P. Behr, F. Zacchello, and M. Scarpa. ExGen 500 is an efficient vector for gene delivery to lung epithelial cells in vitro and in vivo. *Gene Therapy*, 4: 1100–1106, 1997.

[14] Plank, C., K. Mechtler, F.C. Szoka, Jr., and E. Wagner. Activation of the complement system by synthetic DNA complexes: A potential barrier for intravenous gene delivery. *Human Gene Therapy*, 7: 1437–1446, 1996.

[15] Howard, K.A., U.L. Rahbek, X.D. Liu, C.K. Damgaard, S.Z. Glud, M.O. Andersen, M.B. Hovgaard, A. Schmitz, J.R. Nyengaard, F. Besenbacher, and J. Kjems. RNA interference in vitro and in vivo using a chitosan/siRNA nanoparticle system. *Molecular Therapy*, 14: 476–484, 2006.

[16] Pille, J.Y., H. Li, E. Blot, J.R. Bertrand, L.L. Pritchard, P. Opolon, A. Maksimenko, H. Lu, J.P. Vannier, J. Soria, C. Malvy, and C. Soria. Intravenous delivery of antiRhoA small interfering RNA loaded in nanoparticles of chitosan in mice: Safety and efficacy in xenografted aggressive breast cancer. *Human Gene Therapy*, 17: 1019–1026, 2006.

[17] Marques, J.T., T. Devosse, D. Wang, M. Zamanian-Daryoush, P. Serbinowski, R. Hartmann, T. Fujita, M.A. Behlke, and B.R.G. Williams. A structural basis for discriminating between self and nonself double-stranded RNAs in mammalian cells. *Nature Biotechnology*, 24: 559–565, 2006.

[18] Judge, A.D., G. Bola, A.C. Lee, and I. MacLachlan. Design of noninflammatory synthetic siRNA mediating potent gene silencing in vivo. *Molecular Therapy*, 13: 494–505, 2006.

[19] Melnikova, I. RNA-based therapies. *Nature Reviews Drug Discovery*, 6: 863–864, 2007.

[20] DeVincenzo, J., J.E. Cehelsky, R. Alvarez, S. Elbashir, J. Harborth, I. Toudjarska, L. Nechev, V. Murugaiah, A. Van Vliet, A.K. Vaishnaw, and R. Meyers. Evaluation of the safety, tolerability and pharmacokinetics of ALN-RSV01, a novel RNAi antiviral therapeutic directed against respiratory syncytial virus (RSV). *Antiviral Research*, 77: 225–231, 2008.

[21] Davis, M.E., J.E. Zuckerman, C.H. Choi, D. Seligson, A. Tolcher, C.A. Alabi, Y. Yen, J.D. Heidel, and A. Ribas. Evidence of RNAi in humans from systemically administered siRNA via targeted nanoparticles. *Nature* 464: 1067–1070, 2010.

[22] Aleku, M., P. Schulz, O. Keil, A. Santel, U. Schaeper, B. Dieckhoff, O. Janke, J. Endruschat, B. Durieux, N. Roder, K. Loffler, C. Lange, M. Fechtner, K. Mopert, G. Fisch, S. Dames, W. Arnold, K. Jochims, K. Giese, B. Wiedenmann, A. Scholz, and J. Kaufmann. Atu027, a liposomal small interfering RNA formulation targeting protein kinase N3, inhibits cancer progression. *Cancer Research*, 68: 9788–9798, 2008.

7

Chitosan

7.1 Introduction

Chitosan, an aminoglucopyran, is composed of randomly distributed N-acetylglucosamine and glucosamine residues. Owing to its versatile biological activity, biocompatibility, and biodegradability along with low toxicity, chitosan has emerged as a promising candidate of highly sophisticated functions [1]. To harness the unique properties and to tap full potential of this versatile polysaccharide, enormous attempts have been made to functionalize and derivatize it. Chitin is the second most abundant natural biopolymer extracted from exoskeletons of crustaceans and from cell walls of fungi and insects. Structurally, chitin is a linear cationic heteropolymer of randomly distributed N-acetylglucosamine and glucosamine residues with β-1,4-linkage. Chitosan is produced by N-deacetylation of chitin in the presence of alkali (Figure 7.1). Controlled derivatization of chitin results in chitosan with DDA between 40% to 98% and the MW between 5×10^4 Da and 2×10^6 Da [2]. The biological application of chitosan depends on DDA and degree of polymerization (DP), which also decides the MW of polymer. Chitosan possesses reactive hydroxyl and amino groups and is usually less crystalline than chitin. Upon heating, it degrades prior to melting, thus this polymer has no melting point. Chitosan is readily soluble in dilute acidic solutions below pH 6.0. It can be considered as a strong base as it possesses primary amino groups with a pKa value of 6.3. Due to the presence of amino groups, the charged state and properties of chitosan are dictated by the pH [3]. At low pH, the amino groups get protonated and become positively charged, making chitosan a water-soluble

S. Nimesh and R. Chandra,
Theory, Techniques and Applications of Nanotechnology in Gene Silencing, 79–92.
© 2011 *River Publishers. All rights reserved.*

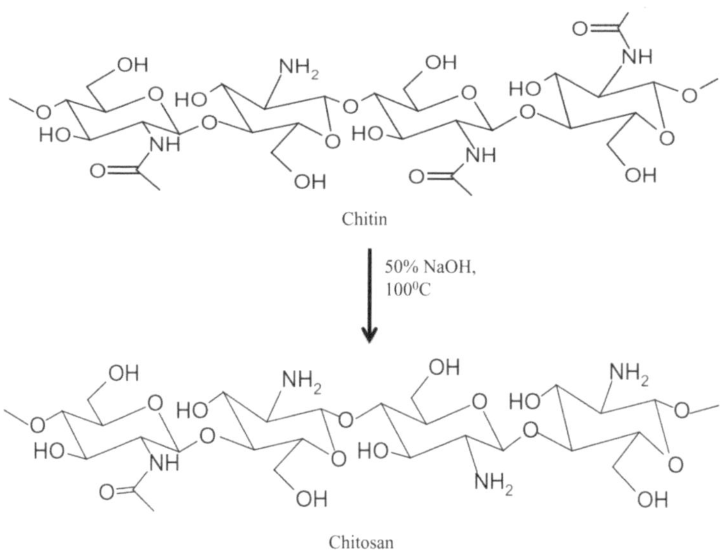

Fig. 7.1 Preparation of chitosan from chitin.

cationic polyelectrolyte. On the other hand, as the pH is increased above 6, chitosan amino groups become deprotonated; the polymer loses its charge and becomes insoluble. This transition between solubility-insolubility occurs at its *pKa* value between pH 6 and 6.5. Under physiological conditions, chitosan can be easily digested either by lysozymes or by chitinases, which can be produced by the normal flora in the human intestine or exist in the blood [4–6]. Owing to these properties, it has been widely employed for drug delivery both in pharmaceutical research and in industry [7]. Lately, chitosan has been investigated as a safer alternative to other non-viral vectors such as cationic lipids for siRNA delivery.

7.2 Parameters Affecting siRNA Delivery Efficiency of Chitosan-based Nanoparticles

Chitosan is one of the most widely investigated non-viral, naturally derived polymeric gene delivery vector. The transfection efficiency of chitosan/DNA nanoparticles depends on several factors such as the DDA and MW of the chitosan, pH, protein interactions, N/P ratio, cell type, nanoparticle size and interactions with cells [8]. The DNA binding affinity and transfection efficiency have been found to increase with an increase in the DDA or MW while maximum expression levels are achieved by obtaining an intermediate stability through control of MW and DDA [9, 10]. Recently, chitosan has been reported to deliver siRNA. The siRNA delivery efficacy of chitosan is dictated by various structural and experimental parameters. There are two methods of generating siRNAs for gene silencing studies. One involves the preparation of siRNAs via chemical or enzymatic synthesis. The other relies on siRNA expression vectors. Synthetic dsRNA, which mimics siRNA, is generated by Dicer and can induce gene silencing processes. However, synthetic siRNA has a short half-life in cells, making its effect transient.

7.2.1 Chitosan Molecular Weight and Concentration

The physicochemical properties i.e. size, zeta potential, morphology, complex stability and *in vitro* gene silencing of chitosan/siRNA nanoparticles is influenced by chitosan MW. Stability of nanoparticles is desirable for extracellular siRNA protection; however, decomplexation is essential for siRNA to mediate gene silencing through interaction with intracellular components such as the RISC. Hence, a subtle balance between protection and release of siRNA is required for biological activity of siRNA delivered with chitosan.

The particle size of the chitosan/siRNA complex is dependent on chitosan MW. Katas *et al.* showed that a smaller mean particle size of chitosan nanoparticles was obtained when the lower MW of chitosan (110 kDa) was used compared to the higher MW (270 kDa) chitosan [11]. Liu *et al.* investigated chitosan/siRNA nanoparticles formulated in MW range 8.9–173 kDa [12]. The size of the particles at low MW (8.9 kDa) was 3500 nm which further reduced to 200nm when chitosan of high MW (173 kDa) was employed. The chitosan molecules (MW 64.8–170 kDa), 5–10 times the length of the siRNA (MW of 13.36 kDa), could form stable complexes with siRNA through electrostatic forces resulting in high gene silencing efficiency in H1299 human lung carcinoma cells. On the contrary, chitosan with a low MW (10 kDa) could not condense siRNA into stable particles, resulting in the formation of large aggregates and almost negligible knockdown. Also, Ji *et al.* observed very small nanoparticles of 148 nm with low polydispersity index using higher MW chitosan (190–310 kDa) [13]. These nanoparticles which incorporated siRNA specific for the FHL2 gene reduced about 70% FHL2 gene expression in human colorectal cancer Lovo cells. Hence, MW of chitosan could be considered as an important parameter dictating its siRNA-mediated knockdown efficiency.

The chitosan concentration also influences the properties of the complexes. Mean particle size of chitosan-siRNA complexes prepared by simple complexation for both chitosan hydrochloride and glutamate increased when the concentration of chitosan was increased from 25 to 300 μg/ml in distilled water [11]. Also, the comparative positive value of surface charge (zeta potential) of the chitosan-siRNA complexes increased with the increasing concentration of chitosan at a constant siRNA concentration.

7.2.2 Degree of Deacetylation

The positive charge density of chitosan depends on the DDA when dissolved in acidic conditions. The DDA value represents the percentage of deacetylated primary amino groups along the molecular chain, which further determines the positive charge density of chitosan. The chitosan with higher DDA (DDA usually above 80%) has been shown to possess better siRNA binding capacity [12]. On the contrary, chitosan with low DDA has low charge interaction with siRNA. Here,the particles formed are unstable and showed low knockdown in H1299 human lung carcinoma cells [12].

7.2.3 N/P ratio

For chitosan/siRNA nanoparticles, the N/P ratio is defined as the ratio of amino groups (N) of chitosan to phosphate groups (P) of siRNA. Howard *et al.* studied the influence of N/P ratio on the size of chitosan/siRNA nanoparticles and reported higher particle size at lower N/P ratio [14]. For instance, at low concentrations (250 μg/ml) of chitosan, nanoparticles formed at N/P ratio of 71 and measured 181.6 nm but increased to 223.6 nm at N/P ratio of 6. Further, Liu *et al.* investigated the influence of N/P ratio on the gene silencing efficacy of chitosan 170 kDa (DDA 84%) in H1299 human lung carcinoma cells and found that the level of EGFP silencing increased at higher N/P ratios (50 and 150) in comparison to low N/P ratio formulations [2 and 10]. Nanoparticles formed at N/P 150 showed the greatest level (80%) of EGFP expression silencing [12]. The total chitosan used at high N/P ratios does not completely participate in the formation of the nanoparticles. It was therefore proposed that this proportion of chitosan is more likely to loosely associate with nanoparticles at high N/P and contribute to the improved stability and increased gene silencing. Moreover, removal of excess chitosan prior to transfection resulted in virtually no cellular knockdown, suggesting the requirement of stabilized particles for cellular entry or a possible role of excess chitosan in cellular permeation.

7.2.4 Chitosan Salt Form

The properties of the chitosan-siRNA nanoparticles are also dictated by chitosan salt form. Chitosan glutamate (G213, G113), which has higher MW than chitosan hydrochloride (Cl213, Cl113) produced smaller complexes with siRNA than chitosan hydrochloride [11]. Complete binding of siRNA with the chitosan was observed when the chitosan nanoparticles to siRNA weight ratio approached 100:1 except for lower MW chitosan hydrochloride, Cl113. Also, higher siRNA loading efficiency was observed for siRNA adsorbed onto chitosan glutamate (83% ± 0.9% for G213, and 90% ± 0.3% for G113) compared to chitosan hydrochloride (72% ± 1.1% for Cl213, and 59% ± 0.8% for Cl113) [11]. Further, chitosan glutamate, G213 (470 kDa), showed the highest gene silencing effect at 24 h post-transfection either by simple complexation (51% of gene knockdown) or ionic gelation (82% and 63% of gene knockdown for siRNA entrapment and siRNA adsorption, respectively) compared to chitosan hydrochloride in CHO K1 cells.

7.2.5 pH of the System

Although, pH of the system determines the chitosan charge, investigation of the influence of pH on the chitosan-siRNA nanoparticles size revealed that no significant difference in particle size was observed when complexing chitosan with siRNA in acetate buffer at pH 4.5 (0.1 M) compared to complexing in distilled water [11]. However, in a recent study by AFM, the interaction strength between siRNA and chitosan was found to be pH dependent where the adhesive interactions decreased as the pH increased from 4.1 to 6.1, 7.4, and 9.5, exhibiting distinct multimodal distributions of the interaction forces between siRNA and chitosan molecules at acidic pH while only negligible adhesive forces were observed at neutral or high pH. The strong pH dependence of siRNA-chitosan interaction can provide a convincing rationale for chitosan-siRNA complex formation and nanoparticle stability under low acidic conditions [15]. The influence of pH on the gene silencing activity of siRNA in cell culture has not been reported so far and exhaustive studies are required to further understand this parameter. We have systematically examined the effect of pH on the physicochemical properties and gene delivery efficacy of chitosan-DNA nanoparticles [16]. Nanoparticles prepared with low MW chitosan (10 kDa) suspended in double distilled water (pH 6.1) gave a uniform hydrodynamic diameter distribution of 243 ± 12 nm which increased to 911 ± 39.6 nm and 1213 ± 84 nm in PBS pH 6.5 and pH 7.1 respectively. Cells incubated for 24 hours with complexes prepared from rhodamine labeled chitosan revealed the presence of large aggregates at pH 7.4 adhering to cell surfaces while RITC-chitosan fluorescence appeared to be more uniformly distributed on the cells at pH 6.5 (Figure 7.2). Transfection efficiency expressed as % of cells expressing EGFP was 26.3% at pH 6.5 and then dropped considerably

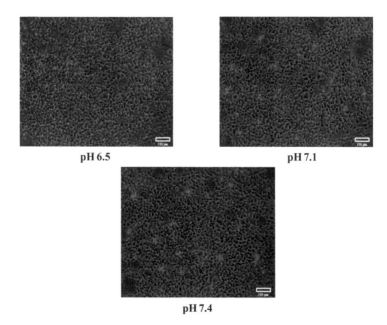

pH 6.5   pH 7.1

pH 7.4

Fig. 7.2 Fluorescence microscope images of HEK 293 cells transfected at different pH. Cells were exposed to rhodamine labeled chitosan/DNA complexes, and 24 h posttransfection media was exchanged with PBS and cells with internalized or adsorbed complexes visualized under fluorescent microscope at 10x magnification. Large aggregates present on cell surface at pH 7.4 while complexes were more uniformly distributed at lower pH 6.5. Adapted from Nimesh *et al.* [16].

at higher pH of 7.1 and 7.4 with 9.2% and 0.2%, respectively [16]. Further, uptake of nanoparticles by HEK 293 cells, as assessed by flow cytometry, was pH dependent with maximum uptake occurring in medium at pH 6.5.

7.2.6 Presence of Serum

Stability against digestion by nucleases is highly desirable for siRNA to achieve maximal knockdown activity in cells. Katas *et al.* investigated the stability of chitosan-siRNA nanoparticles in the presence of serum by incubating free and chitosan complexed siRNA in 5% FBS at 37° C, respectively [11]. It was observed that siRNA was intact only up to 30 min and it was fully degraded after 48 h. On the contrary, siRNA recovered from the chitosan-TPP nanoparticles started to degrade after 24 h incubation and fully degraded after 72 h incubation. The experiment was further extended by incubating siRNA as well as chitosan-siRNA nanoparticles in 50% serum concentration. The chitosan-siRNA nanoparticles remarkably protected siRNA from nuclease activity. Complete degradation of siRNA was observed as early as time-point zero with degradation occurring during the mixing of siRNA with serum and freezing steps. In contrast with unformulated siRNA, the siRNA recovered from chitosan-siRNA nanoparticles was intact up to 7 h and fully degraded after 48 h incubation in 50% serum.

7.2.7 Methods of siRNA Association

The association of siRNA with the chitosan has been observed to play an important role in the silencing effect. Katas *et al.* compared three methods of siRNA association: Simple complexation, ionic gelation (siRNA entrapment) and adsorption of siRNA on the surface of preformed chitosan nanoparticles for preparation of chitosan-based nanoparticles [11]. Owing to their high binding capacity and loading efficiency, chitosan-TPP nanoparticles with entrapped siRNA were shown to be better vectors as siRNA delivery vehicles compared to chitosan-siRNA complexes in two different types of cell lines, CHO K1 and HEK 293. On the other hand, a low inhibition of gene expression was observed for nanoparticles prepared by simple complexation and adsorption of siRNA on the preformed chitosan-tripolyphosphate (TPP) nanoparticles.

7.3 *In Vitro* and *In Vivo* Applications of Chitosan/siRNA Complexes

From the ongoing research, it is evident that chitosan- and chitosan-based vectors are safe and efficient gene delivery systems. But clinical trials are hampered due to their low transfection efficiency. Although, the formation of stable chitosan-siRNA nanoparticles is a prerequisite for stability and extra cellular protection of siRNA, efficient disassembly is needed to allow RNA-mediated gene silencing. Thus, a proper balance between protection and release plays a vital role in the biological activity of siRNA [17].

Katas *et al.* seems to be the first group to report the use of chitosan to deliver siRNA *in vitro* [11]. Nanoparticles of chitosan-TPP entrapping siRNA were found to be better vectors as compared to chitosan-siRNA complexes possibly due to their high binding capacity and loading efficiency. Upon comparing the gene silencing activity at 24 h and 48 h post-transfection, higher activity was obtained at 24 h than 48 h post-transfection for the cells treated with siRNA associated with chitosan nanoparticles or complexes, suggesting that the release of siRNA might occur within the first 24 h for most of the tested chitosans. Howard *et al.* engineered a novel chitosan-based siRNA nanoparticle which mediated knockdown of endogenous enhanced green fluorescent protein (*EGFP*) in both H1299 human lung carcinoma cells and murine peritoneal macrophages (77.9% and 89.3% reduction) [14]. Further, the ability of the nanoparticles to knockdown expression of the BCR/ABL-1 oncogene, found in chronic myelogenous leukemia, was tested to demonstrate the therapeutic potential of the chitosan system. The BCR/ABL-1-expressing cell line K562 was transiently transfected with nanoparticles (N/P 57) complexed with either a fusion-specific or a non-specific siRNA. Western blotting analysis using an antibody recognizing the N-terminal part of BCR, demonstrated ~90% BCR/ABL-1 knockdown in cells transfected with nanoparticles containing BCR/ ABL-1-specific siRNA. A detailed study showed that the physicochemical properties (size, zeta potential, morphology, and complex stability) as well as *in vitro* gene silencing of chitosan/siRNA nanoparticles are dictated by chitosan MW and DDA [12]. High MW and DDA chitosan resulted in the formation of discrete stable nanoparticles ~ 200 nm in size. Chitosan/siRNA formulations (N/P 50) prepared with low MW (~10 kDa) showed almost no knockdown of enhanced green flurescent protein (EGFP) in H1299 cells, whereas those prepared from higher MW (64.8–170 kDa) and

DDA (~80%) showed gene silencing ranging between 45% and 65%. The highest gene silencing (80%) was achieved using chitosan/siRNA nanoparticles at N/P 150 using higher MW (114 and 170 kDa) and DDA (84%) that correlated with formation of stable nanoparticles of ~ 200 nm. On the contrary, Rojanarata *et al.* reported maximal silencing (70–73%) of *EGFP* gene by chitosan-thiamine pyrophosphate complexes prepared from low MW chitosan (20 and 45 kDa) at a weight ratio of 80 [18]. The observed high efficiency was attributed to increased siRNA binding and improved water solubility of chitosan owing to incorporation of extra amino groups from thiamine pyrophosphate, and salt formation between the phosphate group of thiamine pyrophosphate and the amino group of chitosan [18].

Anderson *et al.* described specific and efficient knockdown of *EGFP* (approximately 70%) in H1299 human lung carcinoma cells transfected in plates precoated with a chitosan/siRNA formulation containing sucrose as lyoprotectant [19]. This methodology alleviates the necessity for siRNA complexation immediately prior to use and addition to the cells. Moreover, chitosan/siRNA formulation displayed silencing activity over the period studied (approximately 2 months) when stored at room temperature. Delivery of shRNA employing chitosan nanoparticle have also been reported. Potent knockdown of TGFB1 by shRNA resulting in a decrease in rhabdomyosarcoma (RD) cell growth *in vitro* has been observed [20]. Controlled size chitosan nanoparticles (110 to 430 nm) were prepared using coacervation method in the presence of polyguluronate encapsulating siRNA [21]. These nanoparticles were not only efficient in delivering siRNA to HEK 293FT and HeLa cells but also showed low cytotoxicity. In another study, chitosan/siRNA nanoparticles were observed as irregular, lamellar and dendritic structures with a hydrodynamic radius size of about 148 nm and net positive charge with zeta potential value of 58.5 mV [22]. The knockdown effect of the chitosan/siRNA nanoparticles showed that FHL2 siRNA formulated within chitosan nanoparticles could knockdown about 69.6% FHL2 gene expression.

Although siRNA delivery employing polymers is quite a new concept, still several successful *in vivo* studies have been reported. The gene silencing efficacy of siRNA-polymer complexes is influenced by the route of administration. Significant RNA interference was observed after nasal administration of chitosan/siRNA formulations (37% and 43% reduction) in bronchiole epithelial cells of transgenic EGFP mice [14]. Intraperitoneal administration of chitosan/siRNA nanoparticles targeting *TNF-α* expression in systemic macrophages in mice have been observed to downregulate systemic and local inflammation in arthritis [23]. In another study, chitosan-TPP nanoparticles used to deliver shRNA to inhibit TGFB1 expression reported a decrease in tumorigenicity in nude mice [20]. Glud *et al.* examined the pulmonary gene silencing effect of short interfering locked nucleic acids (siLNAs) targeting EGFP in lung bronchoepithelium. Chitosan/siLNA nanoparticles significantly reduced the EGFP protein expression after intranasal delivery [24]. The neuronal uptake and distribution of a chitosan-siRNA targeting specific muscarinic acetylcholine receptors (mAChRs) was investigated by intrathecal injection in rats [25]. The protein levels of mAChR subtypes in the spinal cord were significantly downregulated by siRNA treatment. In an attempt to make a stable and tumor homing vector, glycol chitosan (GC) and PEI were modified with 5β-cholanic acid and mixed to form self-assembled GC-PEI nanoparticles (GC-PEI NPs), which were further complexed with siRNAs against red fluorescent protein (RFP) expression [26]. These nanoparticles showed considerable inhibition of *RFP* gene expression in RFP/B16F10 bearing mice, due to their higher tumor targeting ability. Tumor-targeted chitosan nanoparticles were further prepared by attaching a RGD moiety to chitosan [27]. RGD-chitosan nanoparticles loaded with siRNA significantly increased selective intratumoral delivery in orthotopic animal models of ovarian cancer. In a recent study, aerosolized chitosan/siRNA nanoparticles used to dose transgenic EGFP mice showed significant EGFP gene silencing (68% reduction compared to mismatch group) [28].

Chitosan has been shown to improve the transfection efficiency of other vectors in siRNA delivery as a coating material. Pille *et al.* demonstrated that intravenous administration of encapsulated antiRhoA siRNA in chitosan-coated polyisohexylcyanoacrylate nanoparticles to nude mice with xenografted aggressive breast cancers can significantly inhibit the tumor growth by at least 90% [29]. Further, a chitosan-coated PLGA nanoparticle has been reported to successfully bind the antisense oligonucleotide and to be taken up by A549 cells after 6h of incubation [30].

7.4 Future Perspectives

Although a plethora of reports have been published on the development of efficient chitosan-based vectors for gene delivery, only limited reports are available regarding the application of chitosan and its derivatives for siRNA delivery. The delivery of siRNA is a multistep process that requires bypassing a series of extra- and intracellular barriers. The vector should consist of three different functional moieties: (i) a cationic chitosan polymer component which efficiently condenses siRNA to form complexes, that facilitates endosomal escape of the complexes after cellular uptake, and provides the effective unpacking of the complexes in the cytoplasmic compartment; (ii) a hydrophilic component, such as PEG, which imparts solubility and stability to the complexes in the biological fluid; (iii) a cell- or tissue-specific ligand for enhanced targetibility toward target cells or tissues and the efficient cellular uptake by receptor mediated endocytosis. However, due to the shorter length and linearity of siRNA compared to DNA, it behaves differently. Studies disclosed that siRNA binds to chitosan in a different manner from that observed with pDNA [14]. Recent studies have also indicated that even for the same chitosan vector, different results can be obtained for siRNA delivery compared to that of DNA delivery. Moreover, the extra vulnerability of siRNA to nuclease degradation may offer additional challenges to chitosan-mediated siRNA transfer. Also, more *in vivo* studies will need to be carried out before moving to clinical trials as the majority of reports published till date have only been conducted *in vitro* and in animal models.

7.5 Conclusions

As evident from the literature, the chitosan-mediated siRNA delivery has enormous potential to evolve as therapeutic modalities, despite the existing problems. The transfection efficiency is dictated by a series of formulation-dependent parameters, such as the MW of chitosan, DDA, N/P ratio, chitosan salt form, siRNA concentration, pH, serum, additives, preparation techniques of chitosan/nucleic acid particles and routes of administration. Taking into account the combined effects of the formulation parameters, it is expected that the complexes formed with high MW and high DDA would be highly stable and may lead to low or delayed transfection. On the contrary, complexes formed with a low MW and low DDA chitosan would not be sufficiently stable to transfect cells efficiently. Hence, there is a possibility that a range of intermediate values of MW and DDA could form complexes of intermediate stability and transfect efficiently. Besides formulation optimization, the structural modification of chitosan or additives is an effective way to improve the stability of the polyplexes in biological fluids, enhance targeted cell delivery and facilitate endolysosomal release of the complex. Overall, the transfection efficiency of chitosan-based delivery systems can be adjusted by playing with formulation related parameters.

References

[1] Rinaudo, M. Chitin and chitosan: Properties and applications. *Progress in Polymer Science*, 31: 603–632, 2006.

[2] Hejazi, R. and M. Amiji. Chitosan-based gastrointestinal delivery systems. *Journal of Controlled Release*, 89: 151–165, 2003.

[3] Yi, H., L.Q. Wu, W.E. Bentley, R. Ghodssi, G.W. Rubloff, J.N. Culver, and G.F. Payne. Biofabrication with chitosan. *Biomacromolecules*, 6: 2881–2894, 2005.

[4] Aiba, S. Studies on chitosan: 4. Lysozymic hydrolysis of partially N-acetylated chitosans. *International Journal of Biological Macromolecules*, 14: 225–228, 1992.

[5] Zhang, H. and S.H. Neau. In vitro degradation of chitosan by bacterial enzymes from rat cecal and colonic contents. *Biomaterials*, 23: 2761–2766, 2002.

[6] Escott, G.M. and D.J. Adams. Chitinase activity in human serum and leukocytes. *Infection and Immunity*, 63: 4770–4773, 1995.

[7] van der Lubben, I.M., J.C. Verhoef, G. Borchard, and H.E. Junginger. Chitosan and its derivatives in mucosal drug and vaccine delivery. *European Journal of Pharmaceutical Sciences*, 14: 201–207, 2001.

[8] Mansouri, S., Y. Cuie, F. Winnik, Q. Shi, P. Lavigne, M. Benderdour, E. Beaumont, and J.C. Fernandes. Characterization of folate-chitosan-DNA nanoparticles for gene therapy. *Biomaterials*, 27: 2060–2065, 2006.

[9] Lavertu, M., S. Methot, N. Tran-Khanh, and M.D. Buschmann. High efficiency gene transfer using chitosan/DNA nanoparticles with specific combinations of molecular weight and degree of deacetylation. *Biomaterials*, 27: 4815–4824, 2006.

[10] Ma, P.L., M. Lavertu, F.M. Winnik, and M.D. Buschmann. New insights into chitosan-DNA interactions using isothermal titration microcalorimetry. *Biomacromolecules*, 10: 1490–1499, 2009.

[11] Katas, H. and H.O. Alpar. Development and characterisation of chitosan nanoparticles for siRNA delivery. *Journal of Controlled Release*, 115: 216–225, 2006.

[12] Liu, X.D., K.A. Howard, M.D. Dong, M.O. Andersen, U.L. Rahbek, M.G. Johnsen, O.C. Hansen, F. Besenbacher, and J. Kjems. The influence of polymeric properties on chitosan/siRNA nanoparticle formulation and gene silencing. *Biomaterials*, 28: 1280–1288, 2007.

[13] Ji, A.M., D. Su, O. Che, W.S. Li, L. Sun, Z.Y. Zhang, B. Yang, and F. Xu. Functional gene silencing mediated by chitosan/siRNA nanocomplexes. *Nanotechnology*, 20: 405103, 2009.

[14] Howard, K.A., U.L. Rahbek, X.D. Liu, C.K. Damgaard, S.Z. Glud, M.O. Andersen, M.B. Hovgaard, A. Schmitz, J.R. Nyengaard, F. Besenbacher, and J. Kjems. RNA interference in vitro and in vivo using a chitosan/siRNA nanoparticle system. *Molecular Therapy*, 14: 476–484, 2006.

[15] Xu, S., M. Dong, X. Liu, K.A. Howard, J. Kjems, and F. Besenbacher. Direct force measurements between siRNA and chitosan molecules using force spectroscopy. *Biophysical Journal*, 93: 952–959, 2007.

[16] Nimesh, S., M.M. Thibault, M. Lavertu, and M.D. Buschmann. Enhanced gene delivery mediated by low molecular weight chitosan/DNA complexes: Effect of pH and serum. *Molecular Biotechnology*, 46: 182–196, 2010.

[17] Kim, T.-H., H.-L. Jiang, D. Jere, I.-K. Park, M.-H. Cho, J.-W. Nah, Y.-J. Choi, T. Akaike, and C.-S. Cho. Chemical modification of chitosan as a gene carrier in vitro and in vivo. *Progress in Polymer Science*, 32: 726–753, 2007.

[18] Rojanarata, T., P. Opanasopit, S. Techaarpornkul, T. Ngawhirunpat, and U. Ruktanonchai. Chitosan-thiamine pyrophosphate as a novel carrier for siRNA delivery. *Pharmaceutical Research*, 25: 2807–2814, 2008.

[19] Andersen, M.Ø., K.A. Howard, S.R. Paludan, F. Besenbacher, and J. Kjems. Delivery of siRNA from lyophilized polymeric surfaces. *Biomaterials*, 29: 506–512, 2008.

[20] Wang, S.L., H.H. Yao, L.L. Guo, L. Dong, S.G. Li, Y.P. Gu, and Z.H. Qin. Selection of optimal sites for TGFB1 gene silencing by chitosan-TPP nanoparticle-mediated delivery of shRNA. *Cancer Genetics and Cytogenetics*, 190: 8–14, 2009.

[21] Lee, D.W., K.S. Yun, H.S. Ban, W. Choe, S.K. Lee, and K.Y. Lee. Preparation and characterization of chitosan/polyguluronate nanoparticles for siRNA delivery. *Journal of Controlled Release*, 139: 146–152, 2009.

[22] Yu, J.H., J.S. Quan, J.T. Kwon, C.X. Xu, B. Sun, H.L. Jiang, J.W. Nah, E.M. Kim, H.J. Jeong, M.H. Cho, and C.S. Cho. Fabrication of a novel core-shell gene delivery system based on a brush-like polycation of alpha, beta-poly (L-aspartate-graft-PEI). *Pharmaceutical Research*, 26: 2152–2163, 2009.

[23] Howard, K.A., S.R. Paludan, M.A. Behlke, F. Besenbacher, B. Deleuran, and J. Kjems. Chitosan/siRNA nanoparticle-mediated TNF-alpha knockdown in peritoneal macrophages for anti-inflammatory treatment in a murine arthritis model. *Molecular Therapy*, 17: 162–168, 2009.

[24] Glud, S.Z., J.B. Bramsen, F. Dagnaes-Hansen, J. Wengel, K.A. Howard, J.R. Nyengaard, and J. Kjems. Naked siLNA-mediated gene silencing of lung bronchoepithelium EGFP expression after intravenous administration. *Oligonucleotides*, 19: 163–168, 2009.

[25] Cai, Y.Q., S.R. Chen, H.D. Han, A.K. Sood, G. Lopez-Berestein, and H.L. Pan. Role of M2, M3, and M4 muscarinic receptor subtypes in the spinal cholinergic control of nociception revealed using siRNA in rats. *Journal of Neurochemistry*, 111: 1000–1010, 2009.

[26] Huh, M.S., S.Y. Lee, S. Park, S. Lee, H. Chung, Y. Choi, Y.K. Oh, J.H. Park, S.Y. Jeong, K. Choi, K. Kim, and I.C. Kwon. Tumor-homing glycol chitosan/polyethylenimine nanoparticles for the systemic delivery of siRNA in tumor-bearing mice. *Journal of Controlled Release*, 144: 134–143.

[27] Han, H.D., L.S. Mangala, J.W. Lee, M.M. Shahzad, H.S. Kim, D. Shen, E.J. Nam, E.M. Mora, R.L. Stone, C. Lu, S.J. Lee, J.W. Roh, A.M. Nick, G. Lopez-Berestein, and A.K. Sood. Targeted gene silencing using RGD-labeled chitosan nanoparticles. *Clinical Cancer Research*, 16: 3910–3922, 2010.

[28] Nielsen, E.J., J.M. Nielsen, D. Becker, A. Karlas, H. Prakash, S.Z. Glud, J. Merrison, F. Besenbacher, T.F. Meyer, J. Kjems, and K.A. Howard. Pulmonary gene silencing in

transgenic EGFP mice using aerosolised chitosan/siRNA nanoparticles. *Pharmaceutical Research*, doi: 10.1007/s11095-010-0255-y.

[29] Pille, J.Y., H. Li, E. Blot, J.R. Bertrand, L.L. Pritchard, P. Opolon, A. Maksimenko, H. Lu, J.P. Vannier, J. Soria, C. Malvy, and C. Soria. Intravenous delivery of anti-RhoA small interfering RNA loaded in nanoparticles of chitosan in mice: safety and efficacy in xenografted aggressive breast cancer. *Human Gene Therapy*, 17: 1019–1026, 2006.

[30] Nafee, N., S. Taetz, M. Schneider, U.F. Schaefer, and C.M. Lehr. Chitosan-coated PLGA nanoparticles for DNA/RNA delivery: Effect of the formulation parameters on complexation and transfection of antisense oligonucleotides. *Nanomedicine*, 3: 173–183, 2007.

8

Polyethylenimine

8.1 Introduction

Polyethylenimine (PEI), often considered as the gold standard of gene transfection, is one of the most widely explored cationic polymers for gene delivery.
Since the first successful application by Behr *et al.* of PEI-mediated oligonucleotide delivery, it has been further derivatized to improve the physicochemical and biological properties of polyplexes [1, 2]. High cation density of PEI
(a positive charge per 43 Da, which is the monomer's MW) also contributes to
the formation of highly condensed particles by interacting with nucleic acids.
Depending on the arrangement of the repeating ethylenimine units, PEI occurs
in branched and linear morphological isomers. Branched polyethylenimine
(BPEI) is synthesized by acid-catalyzed polymerization of aziridine whereas
linear polyethylenimine (LPEI) is prepared via ring opening polymerization
of 2-ethyl-2-oxazoline followed by hydrolysis (Figure 8.1) [3, 4].

LPEI contains secondary amines in its backbone except the terminal primary groups. On the other hand, BPEI contains primary, secondary and tertiary
amino groups at the estimated ratio of 1:2:1 [5]. The different types of amine
groups have different *pKa* values and could be protonated at different levels
at a given pH. This confers PEI with a superior buffering capacity spread over
a wide range of pH. PEI uses the "proton sponge" mechanism to promote the
release of endocytosed polyplexes from the endosomes (Figure 8.2) [6–9].
According to this mechanism, the unprotonated amines with different pKa
values confer a buffering effect over a wide range of pH. This buffering may
protect the DNA from degradation in the endosomal compartment during the

S. Nimesh and R. Chandra,
Theory, Techniques and Applications of Nanotechnology in Gene Silencing, 93–108.
© 2011 *River Publishers. All rights reserved.*

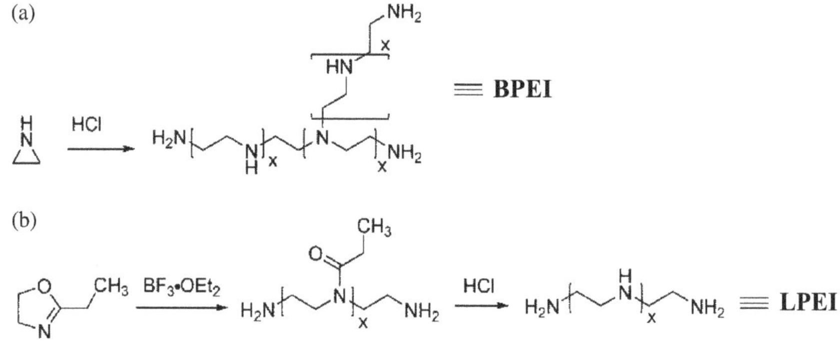

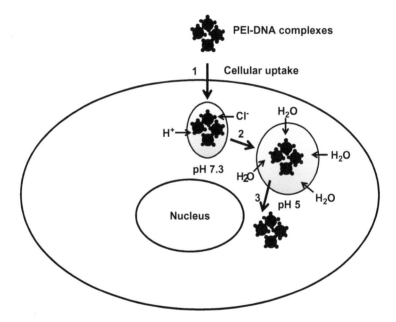

Fig. 8.1 Synthesis of PEI by (A) Acid-Polymerization of Aziridine to yield branched PEI and (B) Ring-Opening polymerization of 2-Ethyl-2-oxazoline followed by hydrolysis to yield linear PEI.

Fig. 8.2 Schematic representation of the "proton sponge effect": the initial step is endocytosis of the cationic complexes (1), followed by acidic endosome buffering (2) which leads to increased osmotic pressure and finally to lysis (3).

maturation of the early endosomes to late endosomes and their subsequent fusion with the lysosomes. The buffering property also allows the polycation PEI to escape from the endosome: At lower pH values, the buffering by PEI causes an influx of chloride ions and water into the endosomes, which eventually burst due to increased osmotic pressure, thus facilitating intracellular release of PEI-DNA polyplexes.

PEIs, ranging from low to high molecular weights, have extensively been exploited as effective gene delivery vehicles [10–14]. PEIs have also been shown to be versatile agents for *in vivo* gene delivery via a number of routes [15–17]. The transfection efficiency of PEI is directly related to size of the polymer and the charge-associated cytotoxicity [11, 18–20]. PEI with high MW exhibited superior transfection efficiency compared to the PEI with low MW. However, the cytotoxicity limits its applications. It has been earlier reported that among the linear and branched PEIs, the latter ones are more toxic and less efficient for transfection, particularly at higher N/P ratios [21]. We addressed the problems associated with high MW branched PEI (750 kDa) in a very simple and effective way by crosslinking with homo-bifunctional PEG [22]. The degree of PEGylation (crosslinking) was varied to control the size of the particles. It was found that PEI-PEG nanoparticles were 5- to 16-fold more efficient as transfecting agents compared to lipofectin and native PEI itself. The toxicity of PEI-PEG nanoparticles was found to reduce considerably in comparison to PEI polymer, as determined by MTT colorimetric assay. Out of the various formulations prepared, PEI-PEG8000 (5% ionic) nanoparticles were found to be the most efficient transfecting agent for *in vitro* transfection (Figure 8.3). The nanoparticles, by virtue of their compactness and small size, increase the physical concentration of DNA on the cell monolayers and enhance the transfection efficiency.

Recently, the application of PEIs as a siRNA delivery vector gained momentum. Although PEI is highly efficient for DNA delivery, its efficacy in siRNA delivery is considered to be far less pronounced [23]. The reduced efficacy of PEI in the later case arises due to the short length of siRNA. The weaker electrostatic interaction between the negatively charged siRNA and the cationic PEI polymer plays crucial role in the dissociation of PEI-siRNA complexes at the anionic cell surface [24]. However, the availability of LPEI and BPEI in a wide range of molecular weights and their implications in various studies to establish the optimal requirements is promising. The most suitable MW of PEI for complexation with DNA is between 5 and 25 kDa, although some groups reported that high (800 kDa) and low (2 kDa) MW PEIs also showed good transfection at their preferred N/P ratios. Hence, an appropriate design of cationic vector having PEI not only caters to the cytotoxicity concern,

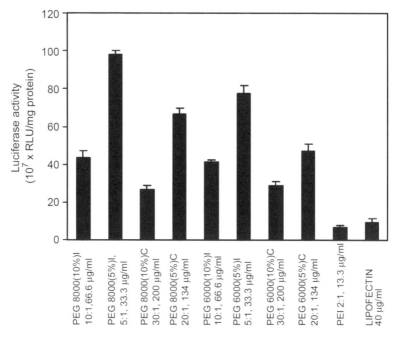

Fig. 8.3 Comparison of transfection efficiency of various PEI-PEG nanoparticle complexes. COS-1 cells were incubated with DNA- PEI-PEG nanoparticle complexes at various weight ratios and incubated for 36 h. The luciferase reporter gene activity in the cell lysate was measured on luminometer and the results are expressed in terms of relative light units/mg total cellular protein. The optimal transfection efficiency obtained by different nanoparticles is represented by bar diagram. The absolute concentration of PEI-PEG nanoparticles complexed with DNA is also indicated. The assays were done in triplicate and the standard error is shown. I-ionic crosslinked, C-covalent crosslinked. Adapted from Nimesh *et al.* [22].

but could also provide sufficient protection against serum or RNase through efficient complexation with siRNA.

8.2   *In Vitro* and *In Vivo* Application of PEI in siRNA Delivery

An appropriate balance between the lipophilicity and hydrophilicity is necessary for the vector to cross the plasma membrane barrier of the cells as well as to reduce the cytotoxicity. However, reduction of the net cationic charge density of the polymers should be moderate and the resulting surface charge must be positive enough to allow the formation of stable polyplexes with siRNA and to exert proton sponge effect to facilitate endosomal release of polyplexes. Hence, there should be a subtle balance between the introduced hydrophobic character and the reduced positive charge. Oskuee *et al.* grafted alkylcarboxylate residues on 25 kDa PEI to investigate the structural influence of the different alkyl modifications on the biophysical properties and bioactivity of the formulation, aiming to improve the efficacy of PEI-based carrier systems for siRNA delivery [25]. Studies on various formulations with varied degree of alkylcarboxylation revealed that at low degree of carboxylation ($< 20\%$), buffering capacity leading to endosomal escape remained efficient. The increase of the hydrophobic alkyl chain length has also been reported to result in improved complex stability with siRNA [26]. The toxicity of modified PEI derivatives reduced significantly in comparison to PEI 25 kDa, owing to the introduction of the additional negative charges lowering the strong interaction potential of PEI with cell surfaces and associated membrane damage. Further, carboxyalkylation of PEI was found to greatly improve siRNA- mediated luciferase gene knockdown.

To overcome instability and low transfection efficiency, PEG (2 kDa)-PEI (25 kDa) was synthesized and investigated as a non-viral carrier of siRNA targeting CD44v6 in gastric carcinoma cells [27]. Physicochemical studies revealed that the size and zeta potential of the complexes below a N/P value 10 were not suitable for cell transfection because PEG-PEI and siRNA could not form a complex of adequate size and positive potential. Further, complete complexation of siRNA was observed with PEG-PEI at N/P ratios ≥ 10. Transfection efficiency of PEG-PEI/siRNA was observed to increase at higher N/P ratios, reaching the maximal transfection efficiency at N/P value 15. However, no major increase appeared at N/P value 30 possibly because over-condensation at high N/P values is not favorable for intracellular release of nucleic acids. Veiseh *et al.* developed the magnetic nanoparticle platform consisting of a superparamagnetic iron oxide (Fe_3O_4) core coated with a cationic copolymer of chitosan-grafted-PEG and PEI for siRNA delivery [28]. To provide site specificity, targeting peptide chlorotoxin was covalently attached to the nanoparticles. This multifunctional magnetic vector was found to deliver siRNA to brain tumor cells through receptor-mediated endocytosis and specifically knockdown the transgene expression of green fluorescent protein (GFP) in C6/GFP + glioma cells. Creusat *et al.* investigated the mechanism of PEI-based siRNA delivery [29]. Results revealed that the underlining proton sponge property is a key to the efficacy of tyrosine-PEI conjugate, as it may allow both endosomal rupture and siRNA liberation via an optimal pH-sensitive dissolution of the PEI self-aggregates. In one of our studies, we prepared an ionic complex of BPEI (750 kDa) and alginic acid to amalgamate the properties of polycationic PEI with polysaccharide alginate [30]. The content of alginic acid was varied systematically to obtain a series of nanocomposites. As the concentration of alginic acid in nanocomposites was increased, the zeta potential and size of nanocomposites decreased accordingly. The zeta potential profile showed that the excess surface charge on PEI (22 mV) is effectively masked in nanocomposites. We observed maximum transfection efficiency in the case of PEI-alginate (6.26%) at w/w ratio 20:1, which was found to be 2 to 16-fold higher than the native PEI on various cell types. Further, these PEI-alginate (6.26%) nanocomposites efficiently delivered siRNAs into mammalian cells, resulting in 80% suppression of the GFP expression (Figure 8.4).

In another published study, we prepared nanoparticles of PEI by acylating PEI with propionic anhydride followed by crosslinking with PEG-bis(phosphate) [31]. The nanoparticles size as revealed by DLS studies was found to be ~110 nm and AFM investigations showed spherical and compact complexes with an average size of 100 nm. The qualitative analysis of the level

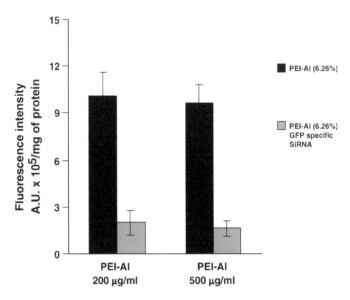

Fig. 8.4 PEI-alginate (6.26%) as an effective carrier of siRNA. PEI-alginate (6.26%) was tested for its ability to deliver GFP specific siRNA into COS-1 cells. The expression of GFP in cell lysates was reduced by 80% as monitored by measuring fluorescence on spectrofluorometer. Adapted from Patnaik *et al.* [30].

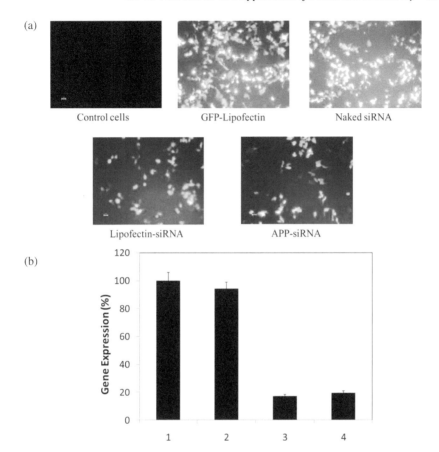

Fig. 8.5 Comparison of gene silencing efficiency of various siRNA formulations after 48 h. (a)The GFP expression was observed under fluorescent microscope at 10X magnification. (b) The level of GFP expression was estimated by quantitation of green fluorescence after 48 h. The data was recorded at optimal inhibition efficiency for APP nanoparticles i.e. APP nanoparticles: siRNA at 30:1 ratio. Adapted from Nimesh *et al.* [31].

of gene silencing was carried out by observing under an inverted microscope after 48 h (Figure 8.5a). *In vitro* siRNA delivery studies in HEK 293 cells with nanoparticles showed up to 85% inhibition of GFP gene expression which was almost equal to that for siRNA-Lipofectin (81% inhibition) (Figure 8.5b).

To integrate low cytotoxicity and higher transfection efficiency, Kim *et al.* developed the water soluble lipopolymer (WSLP) by conjugating the cationic head group of low MW BPEI (1.8 kDa) with a hydrophobic lipid anchor, cholesterol chloroformate [32]. The corresponding complex with siRNA, designed to inhibit human VEGF expression, was investigated as a potential siRNA delivery vehicle. It readily formed nanosized complexes (~100 nm) with siRNA and protected siRNAs from enzymatic degradation in serum conditioned media. WSLP/siRNA complexes transfected in human prostate cancer (PC-3) cells derived from human prostate adenocarcinomas and inhibited the VEGF production significantly, while complexes of WSLP with control siRNA did not show this inhibitory effect. WSLP/siRNA complexes reduced the VEGF production by 40% when compared with the unmodified BPEI. Moreover, WSLP/siRNA complexes reduced tumor volume by 55% at 21 days, and by 65% at 28 days when compared with controls. These results indicate that WSLP has the potential as a siRNA delivering agent and can be applied for antiangiogenic tumor therapy.

In order to develop a multifunctional delivery system which could facilitate target specific treatment of tumor, Biswal *et al.* investigated the folate receptors (FRs)-mediated delivery of dihydrofolate reductase (DHFR) siRNA to silence the DHFR gene in FR-positive KB cells [33]. A DHFR siRNA sequence was cloned into a pSUPER-RNAi vector and complexed with the folate-PEG-PEI (25 kDa) (FOL-PEG-PEI) conjugate. They have carried out the complexation of DHFR siRNA expressing pDNA and FOL-PEG-PEI conjugate, and characterized the FOL-PEG-PEI/pSUPER-siDHFR complex by particle size analyzer, gel retardation, and DNase protection assay. The complex was transfected to (a) FRs overexpressing human epidermal carcinoma (KB) cells, and, (b) FR-negative human lung carcinoma (A549) cells. The transfection studies by fluorescence microscopy and RT-PCR showed that the complex delivered the siRNA vector and inhibited DHFR gene in KB cells; however, it remained unaffected when applied to A549 cells as control. Therefore, FR-mediated delivery of siDHFR complexed with FOL-PEG-PEI conjugate inhibits the DHFR expression in FR positive cells alone. The target specific delivery was further substantiated by the fact that lipofectamine-mediated transfection of pSUPER-siDHFR, delivered the vector and inhibited the DHFR gene in both KB and A549 cells. Target specificity was also confirmed by receptor-blocking studies using free folic acid. This strategy can be extended to a wide range of FR-targeted drug delivery and gene silencing therapeutics by siRNA expression pDNA.

Several targeting moieties such as galactose and pullulan have been utilized in liver-targeted gene delivery owing to their preferential and high binding ability with ASGP-R. Pullulan is a water-soluble polysaccharide consisting of 3 α-1,4-linked glucose polymers with different α-1,6-glucosidic linkages. Kang *et al.* introduced pullulan into PEI for liver targeting [34]. They developed a delivery system of pullulan-containing PEI/siRNA complexes for delivery into mice through the tail vein either by hydrodynamics or nonhydrodynamics-based injection. During systemic injection, the PEI/fluorescein-labeled siRNA complex increased the level of fluorescence in the lung whereas PEI-pullulan/siRNA complex led to an increased fluorescence level in the liver. Further, an increase in N/P ratio of PEI/siRNA complexes resulted in higher mice mortality whereas the introduction of pullulan into PEI dramatically reduced mouse death after systemic injection. Therefore, PEI-pullulan polymeric conjugate provides a useful, low toxic approach to the efficient delivery of siRNA into the liver.

To engineer a stable and tumor-homing nanosized carrier, a new nanosized siRNA carrier system composed of biocompatible/biodegradable glycol chitosan polymer (GC) and PEI was modified with hydrophobic 5β-cholanic acid, and were simply mixed to form self-assembled GC-PEI nanoparticles (GC-PEI) [35]. The freshly prepared GC-PEI nanoparticles showed a stable nanoparticle structure (350 nm) and presented a strong positively charged surface (ζ potential = 23.8) which is enough to complex tightly with negatively charged RFP-siRNAs, designed for inhibiting RFP expression. The siRNA encapsulated nanoparticles (siRNA-GC-PEI) formed compact and stable nanoparticle structures (250 nm) at 1:5 weight ratio of siRNA to GC-PEI nanoparticles. Uptake studies done in red fluorescent protein (RFP) expressing B16F10 tumor cells (RFP/B16F10) showed that the siRNA-GC-PEI nanoparticles displayed a rapid time-dependent cellular uptake profile within 1 h. Moreover, the internalized siRNA-GC-PEI nanoparticles led to specific mRNA breakdowns. Furthermore, the new formulation of siRNA-GC-PEI nanoparticles presented a significant inhibition of RFP gene expression of RFP/B16F10-bearing mice, due to their higher tumor-targeting ability. These results revealed the promising potential of GC-PEI nanoparticles as a stable and effective nanosized siRNA delivery system for cancer treatment.

Lee *et al.* suggested reducible polymerized siRNA as a possible solution for low delivery efficiency of siRNA. Dithiol-modified RFP siRNAs at the 5′-ends of both sense and antisense strands were disulfide-polymerized. Polymerized siRNA (poly-siRNA) was composed of 30% oligomeric siRNA (50~300 bps) and 66% polymeric siRNA (above ~ 300 bps) as fractions, and was reducible in reducing solution through disulfide bond cleavage. Poly-siRNA formed more condensed and nanosized complexes with low MW PEI (1800 Da) by strong electrostatic interaction based on the higher charge density of poly-siRNA, when compared with mono-siRNA. The compact poly-siRNA/PEI complexes prevented the loss and degradation of siRNA from a polyanion competitor and RNases in serum. Furthermore, poly-siRNA/PEI complexes exhibited superior intracellular uptake by murine melanoma cells (B16F10), and was accompanied with RFP gene silencing efficiency of about 80%, compared to untreated cells. These results sufficiently support that strong polyanionic and reducible poly-siRNA can be utilized as a novel powerful therapeutic strategy for human diseases.

Along with tumor models, siRNAs complexed with the linear jetPEI/*in vivo* jetPEI were used for the treatment of other pathologies as well. In a study on the post-exposure protection of guinea pigs against a lethal Ebola virus challenge, the intraperitoneal injection of PEI/siRNA led to targeting of the polymerase (L) gene of the Zaire species of EBOV and protective effects due to a significant reduction in plasma viremia levels [36]. Intraperitoneal application of PEI/siRNA complexes led to the downregulation of hypoxia inducible factor 1α and plasminogen activator inhibitor 1 and to a statistically significant reduction of post-operative abdominal adhesion formation [37]. Ge *et al.* demonstrated that the administration of PEI-complexed siRNAs specific for the conserved regions of influenza virus genes allow the treatment and prevention of lethal influenza infections in the mouse. More specifically, upon administration of the complexes before or after the initiation of the virus infection, reduced virus production in the lungs of infected mice was observed [38]. Concomitantly, the lung was identified as one preferential organ of PEI/siRNA complex delivery upon their intravenous injection [38]. This is mainly attributed to the fact that the lung is reached rather quickly after intravenous injection and that the physical properties of the complexes allow the efficient interaction with the pulmonary vasculature. Additionally, a certain degree of the PEI complex aggregation under physiological conditions may be beneficial for efficient cell transfection in the lung. Intravenous injection was employed for targeting IL-13 as well. PEI-complexed, chemically modified siRNAs were administered to sensitized mice just before airway challenge with allergen. The treatment significantly reduced airway resistance in the sensitized and challenged mice suggesting that PEI/siRNA-mediated IL-13 knockdown may prevent the induction of allergen-induced airway dysfunction [39]. Another study showed that systemic delivery of complexes based on fully deacetylated PEI resulted in a marked improvement of siRNA delivery to the lung as well as in reduced toxicity. This was demonstrated by the downregulation of luciferase as a model gene or of the influenza viral nucleocapsid protein gene with a subsequent 94% drop of virus titers in the lungs of influenza-infected animals [40]. In order to decrease toxicity and enhance bioactivity, Werth *et al.* described the purification of a low MW PEI (PEI-F25-LMW), from the commercially available 25 kDa PEI through size exclusion chromatography [41]. Complexes based on PEI-F25-LMW can be stored lyophilized or frozen, thus avoiding the need to prepare complexes freshly and allowing the preparation of standardized aliquots [41, 42].

8.3 Degradable PEI for siRNA Delivery

Although, high MW PEI (25 kDa) is one of the most potent polymeric vectors because of its high pH-buffering capacity for endosomal escape, it lacks degradable linkages and is too toxic for therapeutic applications. Hence, low MW PEI has been suggested as an alternative to high MW PEI (25 kDa). To overcome the poor transfection efficiency associated with low MW PEI, degradable PEIs consisting of low MW PEIs and degradable cross-linkers for intracellular degradation have been reported. These PEIs displayed high transfection efficiency and low cytotoxicity due to their rapid degradation into small MW water soluble fragments, which are easily processed and removed by the cells. Tarcha *et al.* seem to be the first group to report of degradable PEI based on chemically condensed low MW PEI containing beta-aminopropionamide for siRNA delivery [43]. The polymer was synthesized by N-acylation of degradable PEI prepared by Michael reaction of low MW PEI (800 Da) and hexanediol diacrylate to improve chemical stability relative to ester-containing polymers, but in comparison to PEI (25 kDa), better degradability through the amide linkages. The polymer showed significant *in vitro* knockdown of the luciferase gene, up to 80%, in comparison to nontargeting siRNA in stably transfected HuH7 cells [43]. Breunig *et al.* synthesized degradable PEI by introducing disulfide bonds into the low MW PEI and studied the relationship between cellular uptake and gene silencing activity among linear PEI (5 kDa), crosslinked PEI and branched PEI (25 kDa) [44]. The results indicated that the cellular uptake of siRNA was more efficient with increasing branching of the polymer, whereas the siRNA release was promoted by crosslinked PEI, suggesting that a combination of high branching density and reductively cleavable bonds within the PEI may improve siRNA delivery.

In another study, degradable PEI based on LMW PEI (423 Da) and PEG diacrylate (258 Da) was evaluated for small interfering/small hairpin (si/sh) RNA delivery in A549 cells [45]. The polymer successfully delivered siRNA-targeting EGFP and silenced the EGFP expression. The silencing achieved with the polymer was 1.5-fold higher and safer than native PEI (25 kDa). The polymer also exhibited superior protein kinase Akt1 shRNA delivery, thereby efficiently silenced oncoprotein Akt1. Furthermore, polymer shAkt-mediated Akt1 knockdown hindered cancer-cell growth in A549 cells in an Akt1-specific manner due to the degradability of the polymer. Akt (protein kinase B) is an important regulator of cell survival and plays a key role in cancer by stimulating cell proliferation, inhibiting apoptosis, and modulating the protein translation. Later, the same group studied the Akt1 gene silencing after aerosol delivery of degradable PEI/Akt1 siRNA complexes into *K-ras* and urethane-induced lung-cancer-model mice [46]. The aerosol-delivered Akt1 siRNA suppressed the mRNA and protein expression of Akt1 specifically without affecting the Akt2 and Akt3 in the lungs of *K-ras* mice. The number of tumors and the mean of tumor diameter were significantly decreased by the Akt1 siRNA treatment.

8.4 Future Perspectives

PEI is one of the most efficient gene and siRNA delivery systems. The availability of PEI in a wide MW range enabled it to find applications in several *in vitro* and *in vivo* studies. Although the high MW PEI (25 kDa) is considered as a gold-standard due to its high transfection efficiency, it also causes high toxicity. The formulations with high transfection efficacies and low cytotoxicity will be the key towards the development of siRNA-based clinical therapeutics. Studies employing low MW PEI suffer from low transfection efficiency and those with high MW have high toxicities, hence there is need to find an intermediate MW with benefits of both the worlds. PEI modified with biocompatible neutral polymers or with introduction of biodegradable linkages will be preferred for *in vitro* and *in vivo* siRNA delivery.

8.5   Conclusions

Encouraging results of *in vitro* gene silencing in cell cultures and *in vivo* data in mice models suggest PEI as a promising candidate. The high cationic charge density infers high transfection efficiency to PEI (25 kDa) alongwith high toxicity. The problem of cytotoxicity has been addressed by incorporation of neutral moieties such as the PEG. A new approach that proposes the use of small MW PEI to prepare high MW PEI of desirable MW is fascinating. Initial *in vitro* and *in vivo* studies suggest this approach as the upcoming area in PEI-mediated siRNA therapeutics. In a nutshell, the high transfection efficiency of PEI-based delivery vectors can be maintained by reducing the cytotoxicity of the polymer by introduction of biodegradable linkages.

References

[1] Boussif, O., F. Lezoualc'h, M.A. Zanta, M.D. Mergny, D. Scherman, B. Demeneix, and J.P. Behr. A versatile vector for gene and oligonucleotide transfer into cells in culture and *in vivo*: polyethylenimine. *Proceedings of the National Academy of Sciences of the United States of America*, 92: 7297–7301, 1995.

[2] Neu, M., D. Fischer, and T. Kissel. Recent advances in rational gene transfer vector design based on poly(ethylene imine) and its derivatives. *Journal of Gene Medicine*, 7: 992–1009, 2005.

[3] Jones, G.D., A. Langsjoen, S.M.M.C. Neumann, and J. Zomlefer. The Polymerization of ethylenimine. *Journal of Organic Chemistry*, 09: 125–147, 1944.

[4] Brissault, B., A. Kichler, C. Guis, C. Leborgne, O. Danos and H. Cheradame. Synthesis of linear polyethylenimine derivatives for DNA transfection. *Bioconjugate Chemistry*, 14: 581–587, 2003.

[5] Von Harpe, A., H. Petersen, Y. Li, and T. Kissel. Characterization of commercially available and synthesized polyethylenimines for gene delivery. *Journal of Controlled Release*, 69: 309–322, 2000.

[6] Boussif, O., F. Lezoualc'h, M.A. Zanta, M.D. Mergny, D. Scherman, B. Demeneix, and J.P. Behr. A versatile vector for gene and oligonucleotide transfer into cells in culture and *in vivo*: polyethylenimine. *Proceedings of the National Academy of Sciences of the United States of America*, 92: 7297–7301, 1995.

[7] Thomas, M., J.J. Lu, Q. Ge, C. Zhang, J. Chen, and A.M. Klibanov. Full deacylation of polyethylenimine dramatically boosts its gene delivery efficiency and specificity to mouse lung. *Proceedings of the National Academy of Sciences of the United States of America*, 102: 5679–5684, 2005.

[8] Thomas, M. and A.M. Klibanov. Enhancing polyethylenimine's delivery of plasmid DNA into mammalian cells. *Proceedings of the National Academy of Sciences of the United States of America*, 99: 14640–14645, 2002.

[9] Akinc, A., M. Thomas, A.M. Klibanov, and R. Langer. Exploring polyethylenimine-mediated DNA transfection and the proton sponge hypothesis. *The Journal of Gene Medicine*, 7: 657–663, 2005.

[10] Goula, D., C. Benoist, S. Mantero, G. Merlo, G. Levi, and B.A. Demeneix. Polyethylenimine-based intravenous delivery of transgenes to mouse lung. *Gene Therapy*, 5: 1291–1295, 1998.

[11] Boussif, O., F. Lezoualch, M.A. Zanta, M.D. Mergny, D. Scherman, B. Demeneix, and J.P. Behr. A Versatile vector for gene and oligonucleotide transfer into cells in culture and in-vivo - polyethylenimine. *Proceedings of the National Academy of Sciences of the United States of America*, 92: 7297–7301, 1995.

[12] Abdallah, B., A. Hassan, C. Benoist, D. Goula, J.P. Behr, and B.A. Demeneix. A powerful nonviral vector for *in vivo* gene transfer into the adult mammalian brain: polyethylenimine. *Human Gene Therapy*, 7: 1947–1954, 1996.

[13] Erbacher, P., T. Bettinger, E. Brion, J.L. Coll, C. Plank, J.P. Behr, and J.S. Remy. Genuine DNA/polyethylenimine (PEI) complexes improve transfection properties and cell survival. *Journal of Drug Targeting*, 12: 223–236, 2004.

[14] Kunath, K., A. von Harpe, D. Fischer, H. Petersen, U. Bickel, K. Voigt, and T. Kissel. Low-molecular-weight polyethylenimine as a nonviral vector for DNA delivery: comparison of physicochemical properties, transfection efficiency and *in vivo* distribution with high-molecular-weight polyethylenimine. *Journal of Controlled Release*, 89: 113–125, 2003.

[15] Hong, J.W., J.H. Park, K.M. Huh, H. Chung, I.C. Kwon, and S.Y. Jeong. PEGylated polyethylenimine for *in vivo* local gene delivery based on lipiodolized emulsion system. *Journal of Controlled Release*, 99: 167–176, 2004.

[16] Kichler, A., M. Chillon, C. Leborgne, O. Danos, and B. Frisch. Intranasal gene delivery with a polyethylenimine-PEG conjugate. *Journal of Controlled Release*, 81: 379–388, 2002.

[17] Kichler, A., C. Leborgne, E. Coeytaux, and O. Danos. Polyethylenimine-mediated gene delivery: a mechanistic study. *The Journal of Gene Medicine*, 3: 135–144, 2001.

[18] Fischer, D., T. Bieber, Y. Li, H.P. Elsasser, and T. Kissel. A novel nonviral vector for DNA delivery based on low molecular weight, branched polyethylenimine: effect of molecular weight on transfection efficiency and cytotoxicity. *Pharmaceutical Research*, 16: 1273–1279, 1999.

[19] Godbey, W.T., K.K. Wu, and A.G. Mikos. Size matters: molecular weight affects the efficiency of poly(ethylenimine) as a gene delivery vehicle. *Journal of Biomedical Materials Research*, 45: 268–275, 1999.

[20] Ogris, M., S. Brunner, S. Schuller, R. Kircheis, and E. Wagner. PEGylated DNA/transferrin-PEI complexes: reduced interaction with blood components, extended circulation in blood and potential for systemic gene delivery. *Gene Therapy*, 6: 595–605, 1999.

[21] Wightman, L., R. Kircheis, V. Rossler, S. Carotta, R. Ruzicka, M. Kursa, and E. Wagner. Different behavior of branched and linear polyethylenimine for gene delivery *in vitro* and *in vivo*. The *Journal of Gene Medicine*, 3: 362–372, 2001.

[22] Nimesh, S., A. Goyal, V. Pawar, S. Jayaraman, P. Kumar, R. Chandra, Y. Singh, and K.C. Gupta. Polyethylenimine nanoparticles as efficient transfecting agents for mammalian cells. *Journal of Controlled Release*, 110: 457–468, 2006.

[23] Grayson, A.C., A.M. Doody, and D. Putnam. Biophysical and structural characterization of polyethylenimine-mediated siRNA delivery *in vitro*. *Pharmaceutical Research*, 23: 1868–1876, 2006.

[24] Spagnou, S., A.D. Miller, and M. Keller. Lipidic carriers of siRNA: Differences in the formulation, cellular uptake, and delivery with plasmid DNA. *Biochemistry*, 43: 13348–13356, 2004.

[25] Oskuee, R.K., A. Philipp, A. Dehshahri, E. Wagner, and M. Ramezani. The impact of carboxyalkylation of branched polyethylenimine on effectiveness in small interfering RNA delivery. *Journal of Gene Medicine*, 12: 729–738, 2010.

[26] Philipp, A., X. Zhao, P. Tarcha, E. Wagner, and A. Zintchenko. Hydrophobically modified oligoethylenimines as highly efficient transfection agents for siRNA delivery. *Bioconjugate Chemistry*, 20: 2055–2061, 2009.

[27] Wu, Y., W. Wang, Y. Chen, K. Huang, X. Shuai, Q. Chen, X. Li, and G. Lian. The investigation of polymer-siRNA nanoparticle for gene therapy of gastric cancer *in vitro*. *International Journal of Nanomedicine*, 5: 129–136, 2010.

[28] Veiseh, O., F.M. Kievit, C. Fang, N. Mu, S. Jana, M.C. Leung, H. Mok, R.G. Ellenbogen, J.O. Park, and M. Zhang. Chlorotoxin bound magnetic nanovector tailored for cancer cell targeting, imaging, and siRNA delivery. *Biomaterials*, 31: 8032–8042, 2010.

[29] Creusat, G., A.S. Rinaldi, E. Weiss, R. Elbaghdadi, J.S. Remy, R. Mulherkar, and G. Zuber. Proton sponge trick for pH-sensitive disassembly of polyethylenimine-based siRNA delivery systems. *Bioconjugate Chemistry*, 21: 994–1002, 2010.

[30] Patnaik, S., A. Aggarwal, S. Nimesh, A. Goel, A. Ganguli, N. Saini, Y. Singh, and K.C. Gupta. PEI-alginate nanocomposites as efficient *in vitro* gene transfection agents. *Journal of Controlled Release*, 114: 398–409, 2006.

[31] Nimesh, S. and R. Chandra. Polyethylenimine nanoparticles as an efficient *in vitro* siRNA delivery system. *European Journal of Pharmaceutics and Biopharmaceutics*, 73: 43–49, 2009.

[32] Kim, W.J., C.-W. Chang, M. Lee, and S.W. Kim. Efficient siRNA delivery using water soluble lipopolymer for antiangiogenic gene therapy. *Journal of Controlled Release*, 118: 357–363, 2007.

[33] Biswal, B., N. Debata, and R. Verma. Development of a targeted siRNA delivery system using FOL-PEG-PEI conjugate. *Molecular Biology Reports*, 37: 2919–2926, 2010.

[34] Kang, J.-H., Y. Tachibana, W. Kamata, A. Mahara, M. Harada-Shiba, and T. Yamaoka. Liver-targeted siRNA delivery by polyethylenimine (PEI)-pullulan carrier. *Bioorganic and Medicinal Chemistry*, 18: 3946–3950, 2010.

[35] Huh, M.S., S.-Y. Lee, S. Park, S. Lee, H. Chung, S. Lee, Y. Choi, Y.-K. Oh, J.H. Park, S.Y. Jeong, K. Choi, K. Kim, and I.C. Kwon. Tumor-homing glycol chitosan/ polyethylenimine nanoparticles for the systemic delivery of siRNA in tumor-bearing mice. *Journal of Controlled Release*, 144: 134–143, 2010.

[36] Geisbert, T.W., L.E. Hensley, E. Kagan, E.Z. Yu, J.B. Geisbert, K. Daddario-DiCaprio, E.A. Fritz, P.B. Jahrling, K. McClintock, J.R. Phelps, A.C. Lee, A. Judge, L.B. Jeffs, and I. MacLachlan. Postexposure protection of guinea pigs against a lethal ebola virus challenge is conferred by RNA interference. *International Journal of Infectious Diseases*, 193: 1650–1657, 2006.

[37] Segura, T., H. Schmokel, and J.A. Hubbell. RNA interference targeting hypoxia inducible factor 1alpha reduces postoperative adhesions in rats. *Journal of Surgical Research*, 141: 162–170, 2007.

[38] Ge, Q., L. Filip, A. Bai, T. Nguyen, H.N. Eisen, and J. Chen. Inhibition of influenza virus production in virus-infected mice by RNA interference. *Proceedings of the National Academy of Sciences of the United States of America*, 101: 8676–8681, 2004.

[39] Lively, T.N., K. Kossen, A. Balhorn, T. Koya, S. Zinnen, K. Takeda, J.J. Lucas, B. Polisky, I.M. Richards, and E.W. Gelfand. Effect of chemically modified IL-13 short interfering RNA on development of airway hyperresponsiveness in mice. *Journal of Allergy and Clinical Immunology*, 121: 88–94, 2008.

[40] Thomas, M., J.J. Lu, Q. Ge, C. Zhang, J. Chen, and A.M. Klibanov. Full deacylation of polyethylenimine dramatically boosts its gene delivery efficiency and specificity to mouse lung. *Proceedings of the National Academy of Sciences of the United States of America*, 102: 5679–5684, 2005.

[41] Werth, S., B. Urban-Klein, L. Dai, S. Hobel, M. Grzelinski, U. Bakowsky, F. Czubayko, and A. Aigner. A low molecular weight fraction of polyethylenimine (PEI) displays increased transfection efficiency of DNA and siRNA in fresh or lyophilized complexes. *Journal of Controlled Release*, 112: 257–270, 2006.

[42] Hobel, S., R. Prinz, A. Malek, B. Urban-Klein, J. Sitterberg, U. Bakowsky, F. Czubayko, and A. Aigner. Polyethylenimine PEI F25-LMW allows the long-term storage of frozen complexes as fully active reagents in siRNA-mediated gene targeting and DNA delivery. *European Journal of Pharmaceutics and Biopharmaceutics*, 70: 29–41, 2008.

[43] Tarcha, P.J., J. Pelisek, T. Merdan, J. Waters, K. Cheung, K. von Gersdorff, C. Culmsee, and E. Wagner. Synthesis and characterization of chemically condensed oligoethylenimine containing beta-aminopropionamide linkages for siRNA delivery. *Biomaterials*, 28: 3731–3740, 2007.

[44] Breunig, M., C. Hozsa, U. Lungwitz, K. Watanabe, I. Umeda, H. Kato, and A. Goepferich. Mechanistic investigation of poly(ethylene imine)-based siRNA delivery: disulfide bonds boost intracellular release of the cargo. *Journal of Controlled Release*, 130: 57–63, 2008.

[45] Jere, D., C.X. Xu, R. Arote, C.H. Yun, M.H. Cho, and C.S. Cho. Poly(beta-amino ester) as a carrier for si/shRNA delivery in lung cancer cells. *Biomaterials*, 29: 2535–2547, 2008.

[46] Xu, C.X., D. Jere, H. Jin, S.H. Chang, Y.S. Chung, J.Y. Shin, J.E. Kim, S.J. Park, Y.H. Lee, C.H. Chae, K.H. Lee, G.R. Beck, Jr., C.S. Cho, and M.H. Cho. Poly(ester amine)-mediated, aerosol-delivered Akt1 small interfering RNA suppresses lung tumorigenesis. *American Journal of Respiratory and Critical Care Medicine*, 178: 60–73, 2008.

9

Poly-L-lysine

9.1 Introduction

Poly-L-lysine (PLL) is one of the first polymers investigated for non-viral gene delivery, followed by utilization of a large variety of polymers with different MW in physicochemical and biological experiments. It is a cationic polypeptide with amino acid lysine as a repeat unit (Figure 9.1). The DP of lysine can range between 90–450 and lead to the formation of a polypeptide chain with an acceptable degree of biodegradability, a property highly desirable for *in vivo* use. However, the DP has shown to be directly related to toxic effects, where the longer is the lysine chain, the more cytotoxic the PLL would be [1, 2]. Since, PLL of high MW are degraded slowly hence, are toxic to cultured cells [3, 4]. To circumvent the associated cytotoxicity, PLLs can conjugate with coating elements such as PEG or imidazole groups, to balance the cationic charge density and the capability of the PLLs to bind and condense DNA [3, 5]. Further, PLL polypeptides can be also conjugated with other functional elements such as cell ligands to enhance receptor-mediated uptake.

PLL polyplexes are taken up efficiently by the cells as PEI complexes; however, transfection levels remain several orders of magnitude lower. A suggested reason for this is the lack of amino groups with a pKa between 5 and 7, thus allowing no endosomolysis and low levels of transgene expression. Grafting of various targeting moieties or addition of endosomolytic agents like chloroquine or fusogenic peptides have been reported to improve reporter gene expression [6, 7]. Attachment of histidine or other imidazole containing

S. Nimesh and R. Chandra,
Theory, Techniques and Applications of Nanotechnology in Gene Silencing, 109–118.

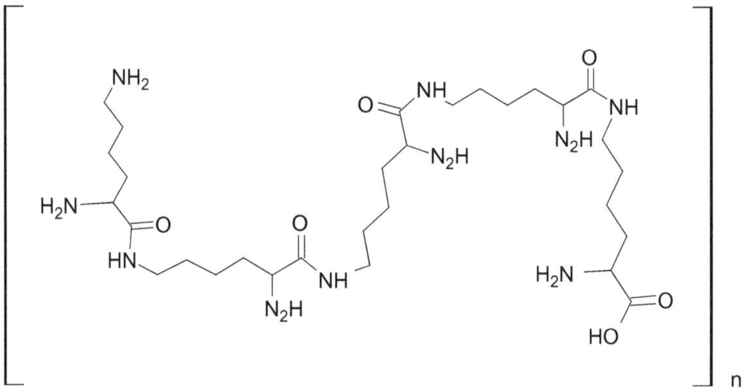

Fig. 9.1 Chemical structure of Poly-L-lysine.

structures to PLL (i.e. pKa around 6 possesses a buffering capacity in the endolysosomal pH range) showed a significant enhancement of reporter gene expression compared to unmodified PLL [8]. High MW PLL (up to 10^6 Da) is synthesized by polymerization of N-carboxy-(N$^\varepsilon$-benzyloxycarbony1)-L-lysine anhydride (Z-L-lysine NCA) [9]. However, controlled MW PLL is prepared by polymerization of N-Carboxy-(F-benzyloxycarbony1)-L-lysine anhydride (Z-L-lysine NCA) in dimethylformamide with triethylamine, diethylamine, or hexylamine as the initiator, at varying molar ratios of NCA to initiator (M/I ratio) [10]. Triethylamine in dioxane gives polypeptides with high MW, whereas primary amines or more polar solvents such as dimethylformamide favor formation of polypeptides of lower MW.

9.2 *In Vitro* and *In Vivo* Applications of PLL in siRNA Delivery

Although PLL has been around for quite a while, it found limited applications in siRNA delivery. PLL presented in different forms e.g. dendrimer, copolymer etc, has been employed to deliver siRNA. Inoue *et al.* investigated the potential of KG6 (6th generation dendritic PLL with 128 amine groups on its surface) to be an efficient siRNA carrier [11]. KG6 was found to deliver fluorescein-labeled oligonucleotide into cells with high efficiency; however, a large amount of the fluorescence was localized in the endosomal compartment, suggesting that KG6-fluorescein-labeled oligonucleotide complexes are unable to move out of the endosomal compartment after endocytic uptake. In order to resolve this issue for KG6, it was combined with Endo-Porter (KG6/EP), which resulted in widespread fluorescence in the cytosol. KG6 also showed significant uptake efficiency (> 90%), in fact, it was the most efficient in contrast to HiPerFect, DoFect-GT1, Lipofectamine 2000, EP, and KG6/EP. Further, KG6 showed effective knockdown of glyceraldehydes-3-phosphate dehydrogenase (GAPDH) with low cytotoxicity in combination with the weak-base amphiphilic peptide —Endo-Porter. In addition, the knock-down of PEPCK, which is the rate limiting enzyme for gluconeogenesis, led to a reduction in glucose production in rat hepatoma H4IIEC3 cells. Knockdown of organic cation transporter 1 (OCT1), which is thought to be the gene that influences metformin action, was shown to successfully diminish the ability of metformin to inhibit gluconeogenesis in H4IIEC3 cells. Later, KG6 was used to deliver ApoB-specific siRNA *in vivo* for the treatment of hypercholes-terolemia [12]. The particle size of siRNA complex of KG6 at cation/anion ratio of 8.0 formed in 5% dextrose as determined by DLS was 168 ± 9.9 nm. The ability of KG6 to deliver siRNA was investigated in C57BL/6 mice by intravenous injection to silence ApoB expression *in vivo*. The ApoB mRNA levels in mice treated with the si-ApoBI complex were reduced by approximately 22% compared with those in the no-treatment group, at all the dosages evaluated. The si-ApoBII complex showed significant reduction of the mRNA — a 50% reduction was observed at 2.5 mg/kg dose. Finally, improvement of hypercholesterolemia was observed after treating ApoE-deficient (ApoE$^{-/-}$) mice with the KG6 siRNA complex. A single intravenous dose of the si-ApoBI complex resulted in a decrease in VLDL and LDL cholesterol levels for up to 96 h, whereas no decrease in their levels was observed with the 5% dextrose-only or the si-Luc complex. The ApoB mRNA levels were also deter-mined 96 h after injection. It was found that the ApoB mRNA levels remained reduced only in the si-ApoBI-treated group (reduction in ApoB mRNA levels: 24.6 ± 2.6% versus 7.6 ± 9.4%, respectively). These results indicate that the systemic siRNA delivery with KG6 could provide a clinically useful approach to reduce cholesterol levels in patients with hypercholesterolemia.

Matsumoto *et al.* prepared a core-shell-type polyion complex (PIC) micelle with a disulfide crosslinked core through the assembly of iminothiolane-modified PEG-block-PLL [PEG-b-(PLL-IM)] and siRNA at the optimum mix-ing ratio [13]. The PIC micelles showed a spherical shape of 60 nm in diameter with a narrow size distribution. The micellar structure was maintained at physiological ionic strength but was disrupted under reductive conditions because of the cleavage of disulfide crosslinks, which is desirable for siRNA release in the intracellular reductive environment. Importantly, environment responsive PIC micelles achieved 100-fold higher siRNA transfection efficacy for siRNA against reporter gene luciferase, compared with non-crosslinked PICs prepared from PEG-b-PLL, which were unstable at physiological ionic strength. PICs formed with PEG-b-(PLL-IM) at non-optimum ratios did not assemble into micellar structure and did not achieve gene silencing following siRNA transfection. These findings show the feasibility of core crosslinked PIC micelles as carriers for therapeutic siRNA and show that stable micellar structure is critical for effective siRNA delivery into target cells.

To overcome the risk of polyplex dissociation in the extracellular environment, siRNA was covalently incorporated into a pH- and redox-responsive polymer conjugate [14]. The novel siRNA conjugate consists of PLL (MW = 32 kDa, DP = 153) as the RNA binding and protecting polycation, PEG as the solubilizing and shielding polymer, the lytic peptide melittin masked by dimethylmaleic anhydride (DMMAn) removable at endosomal pH, and the siRNA attached at the 5′-end of the sense strand via a bioreducible disulfide bond. The PEG-PLL-DMMAn-Mel-siRNA conjugate formed particles in the range of 80–300 nm, depending on the handling (e.g., freeze thawing, dilution).The purified siRNA conjugate was stable in the presence of the polyanion heparin at conditions where the analogous electrostatic siRNA polyplexes disassemble. Only the combination of heparin plus a reducing agent such as glutathione triggered the release of siRNA from the conjugate. The biological efficacy of the luciferase siRNA conjugate as well as viability treated cells was evaluated using Neuro2A-eGFPLuc-cells. High *in vitro* biocompatibility (absence of cytotoxicity or hemolytic activity at neutral pH) and efficient sequence-specific gene silencing was found at ≥ 25 nM siRNA. With PEG-PLL-DMMAn-Mel/luciferase siRNA polyplex (complexed at w/w ratio 2) a 90% knockdown of luciferase expression was observed. The conjugate containing covalently-linked GL3 luciferase siRNA enabled 80% luciferase knockdown at lower siRNA doses (0.125 and 0.25 μg) and a 90% knockdown at higher siRNA amounts (≥0.5 μg). *In vivo* toxicity studies of this formulation demonstrated that conjugates remain to be optimized for therapeutic application.

Lipid modified polymers have been proposed as efficient gene carriers as they condense relatively large plasmid DNAs (>1000 base pairs) and protect them against nuclease degradation [15]. Lipid substitution on cationic polymers enhanced the transport of plasmid DNA across cellular membranes. With these considerations Abbasi *et al.* explored the feasibility of stearic-acid substituted PLL (PLL-StA) as carriers for siRNA delivery with the purpose of P-gp downregulation in a tumor cell model [16]. P-gp is a member of the ABC-transporter family that is responsible for multidrug resistance (MDR) in breast and ovarian cancer cells. PLL-StA vectors protected the siRNA against degradation at low siRNA:carrier ratios (1:3 and 1:1). The lipid modification of PLL led to greater delivery of siRNA into MDR1 cells. The PLL-StA enabled delivery of siRNA to almost all cells (>90% by flow cytometry) but produced a significant drop in siRNA levels after 48 hours. With multiple siRNA treatments over 24-hour intervals, ∼ 40% to 50% P-gp suppression was achieved after 72 hours using PLL-StA and Lipofectamine 2000. The siRNA delivery by PLL-StA resulted in an increase upto 2.5-fold in DOX-positive MDR1 cells. Complementary cytotoxicity studies with the MTT assay revealed increases of ∼35% and ∼25% in the cytotoxicity of DOX-treated cells and PTX-treated cells, respectively, after siRNA delivery by PLL-StA. The authors concluded that effective siRNA delivery with non-viral carriers can reduce the level of P-gp on cell surfaces and enhance the efficiency of chemotherapeutic agents *in vitro.*

The outstanding biocompatibility of PEG, including water solubility, enzymatic tolerance, and minimized non-specific interaction with plasma protein and blood cells makes it a potential candidate for biomedical applications with cationic polymers. PEG could be introduced either by grafting or as a copolymer onto the cationic polymers. Shimizu *et al.* developed a system to target the delivery of siRNAs to glomeruli via PEG-PLL-based vehicles [17]. The size of siRNA/nanocarrier complex was approximately 10–20 nm in diameter that would allow it to move across the fenestrated endothelium to access to the mesangium. siRNAs were detected in the blood circulation for a prolonged time, after intraperitoneal injection of fluorescence-labeled siRNA/nanocarrier complexes. Repeated intraperitoneal administration of a mitogen-activated protein kinase 1 (MAPK1) siRNA/nanocarrier complex suppressed glomerular MAPK1 mRNA and protein expression in a mouse model of glomerulonephritis; this improved kidney function, reduced proteinuria, and ameliorated glomerular sclerosis. Furthermore, this therapy reduced the expression of the profibrotic markers TGF-β1, plasminogen activator inhibitor-1, and fibronectin. Hence, intraglomerular genes were successfully silenced with siRNA using PEG-PLL nanocarriers.

In another study, Sato *et al.* prepared and evaluated a series of cationic comb-type copolymers (CCCs) possessing a polycationic backbone (less than 30 weight (wt) %) and abundant water-soluble side chains (more than 70 wt.%) as a siRNA carrier with prolonged blood circulation time [18]. Markedly, the CCC with the higher side chain content (10 wt.% PLL and 90 wt.% PEG) showed stronger interaction with siRNA than with the lower content (30 wt.% PLL and 70 wt.% PEG), suggesting that highly dense PEG brush reinforces interpolyelectrolyte complex between the PLL backbone and siRNA. The siRNA complexed with the CCC was resistant to nucleases in 90% plasma for 24 h *in vitro*. The CCC having the higher side chain content increased circulation time of siRNA in mouse bloodstream by 100-fold. Surprisingly, even when the CCC and siRNA were separately injected into a mouse at 20 min interval, blood circulation of postinjected siRNA was significantly increased. These results imply that the CCC has higher selectivity in its ionic interaction with siRNA than other anionic substances in blood stream. It appears that this is the first example of a polyplex carrier that prolongs blood circulation time of unmodified siRNA without resource-consuming preparation process. Later, the same group prepared a new series of PEG-grafted PLL (PLL-g-PEG) with various lengths (PEG 2 kDa, 5 kDa, and 10 kDa and PLL 28 kDa and 40 kDa) to evaluate masking effects of PEG on cationic charges of PLL *in vivo* and the structural implications for biodistribution and tumoral accumulation [19]. The best in the series, 40K10P37 (40 kDa of PLL, 10 kDa of PEG, 37 mol% grafting) with MW of 10^6 as determined by multi-angle laser light scattering (MALLS), accumulated in tumors at about 8% of the injected dose per gram of tissue. Interestingly, a PLL-g-PEG conjugate pre-mixed with murine sera prevented degradation of siRNA, suggesting that PLL-g-PEG preferentially associates with siRNA in sera. The results indicate that grafting of PEG to the side chains of PLL augments its lifetime in blood circulation and tumoral accumulation without losing its ability to associate with siRNA and support further evaluation of these cationic delivery carriers.

The endogenous protein albumin is a suitable material for preparation of nanoparticles as it is easily available in large quantities and metabolizes *in vivo* to produce innocuous degradation products. To understand the stability of albumin nanoparticles in an aggressive proteolytic environment, bovine serum albumin (BSA) nanoparticles were fabricated via a coacervation technique and stabilized by coating using different molecular weights (MWs: 0.9–24 kDa) and concentrations (0.1–1.0 mg ml^{-1}) of the cationic polymer, PLL [20]. The nanoparticles thus produced were characterized for morphology (with AFM), size (with PCS) and charge (zeta potential). The size range of engineered BSA particles (155 $\pm$ 11 to 3800 $\pm$ 1600 nm) was effectively controlled by the MW and concentration of the PLL was used for coating. The aqueous solution stability of nanoparticles increased with an increased MW and PLL concentration. However, in the presence of trypsin, nanoparticles coated with higher MW PLL were not as stable as those formed using lower MW PLL. This trend was also confirmed based on the release pattern of siRNA in the presence of trypsin. It was concluded that, while designing stabilizing coatings for soft protein-based nanoparticles, smaller molecules may be more suitable for particle coating if enhanced proteolytic resistance and more stable nanoparticles are desired for targeted drug delivery applications.

High loading of siRNA can be achieved by coating multiple layers of siRNA onto a nanoparticle surface. The enzyme-assisted release of siRNA can be controlled by the number of layers and the degradability of the positively-charged polymers. Moreover, the shielding layers also provide protection to the siRNA from enzymatic degradation. Recently, gold nanoparticles (AuNPs) were employed as core owing to their unique properties, including uniform size, shape-dependent optical and electronic features, biocompatibility, and ease of surface modification, while PLL was employed for layer by layer coating [21]. PLL and siRNA were formulated onto AuNPs, by alternating the charged polyelectrolytes. Up to four layers of PLL and three layers of siRNA (sR3P) are coated. The size of initial bare AuNPs was 40 nm, while the particle size increased steadily with the number of layers (sR1: 104 nm; sR1P: 151 nm; sR2P: 159 nm; sR3P:183 nm). In contrast, TEM images of bare AuNPs (40 nm) and polyelectrolyte-coated AuNPs (sR1P, sR2P, and sR3P) were found to be $\approx$ 50 nm in diameter. The difference between DLS and TEM was argued to be caused by the hydrodynamic structure of sRAuNPs. The release kinetics of siRNA from sRAuNPs depended on the number of layers (sR1P > sR2P > sR3P). The release was slow and took about three days for siRNA to be fully released from the sR1P particles, which had one layer of siRNA and two layers of PLL under the testing condition, whereas it required four and five days for sR2P and sR3P, which had two siRNA/three PLL and three siRNA/four PLL, respectively. The investigation of the ability of sRAuNPs to enter MDA-MB231-luc2 and LNCaPluc2 cell lines revealed that sRAuNPs require no transfection agent to enter cells, and once internalized the siRNA could be freed from particles slowly. Moreover, no significant toxicity was detected for all sRAuNPs in both cell lines. Due to the slow degradation of PLL, the incorporated siRNA was released gradually and showed extended gene silencing effects. Importantly, the inhibition effect in cells is found to correlate with the number of siRNA layers with trilayer siRNA-coated (sR3P) AuNPs; the best formulation in gene silencing (>80%) among all the different kinds of siRNA delivery formulations. Hence, prepared multilayered sRAuNPs, with the outer surface layer of PLL, could deliver siRNA into tumor cells and silence its target gene effectively and without toxicity.

9.3   Future Perspectives

Although PLL is one of the first investigated cationic vectors for gene delivery, the siRNA delivery aspect has made slower progress. Its polyplexes are taken up into cells as efficiently as PEI complexes, however transfection efficiencies remain several orders of magnitude lower. A potential reason for this is the lack of amino groups with a pKa between 5 and 7, thus allowing no endosomolysis and low levels of transgene expression. The manipulations involving fabrication of targeting moieties or co-application of endosomolytic agents like chloroquine or fusogenic peptides will enhance siRNA efficacies. Proper studies to decipher the mechanism of siRNA delivery employing PLL needs to be done along with studies *in vivo* animal models to establish the potential therapeutic relevance.

9.4   Conclusion

The initial encouraging results of siRNA delivery using PLL appears to be promising towards development of safer and effective delivery vectors. The availability of PLL in a wide variety of MW and the degree of polymerization has added advantage as smaller MW PLL may be more suitable for particle coating if enhanced proteolytic resistance and more stable nanoparticles are desired for targeted drug delivery applications. Manipulations involving formulation optimization, the structural modification of PLL or copolymers appears to be efficient way to improve the stability of the polyplex in *in vivo* milieu alongwith enhanced targeted cellular delivery.

References

[1] Plank, C., M.X. Tang, A.R. Wolfe, and F.C. Szoka, Jr. Branched cationic peptides for gene delivery: Role of type and number of cationic residues in formation and *in vitro* activity of DNA polyplexes. *Human Gene Therapy*, 10: 319–332, 1999.

[2] Martin, M.E. and K.G. Rice. Peptide-guided gene delivery. *AAPS Journal*, 9: E18–29, 2007.

[3] Putnam, D., C.A. Gentry, D.W. Pack, and R. Langer. Polymer-based gene delivery with low cytotoxicity by a unique balance of side-chain termini. *Proceedings of the National Academy of Sciences of the United States of America*, 98: 1200–1205, 2001.

[4] Symonds, P., J.C. Murray, A.C. Hunter, G. Debska, A. Szewczyk, and S.M. Moghimi. Low and high molecular weight poly(L-lysine)s/poly(L-lysine)-DNA complexes initiate mitochondrial-mediated apoptosis differently. *FEBS Letters*, 579: 6191–6198, 2005.

[5] Choi, Y.H., F. Liu, J.S. Kim, Y.K. Choi, J.S. Park, and S.W. Kim. Polyethylene glycol-grafted poly-L-lysine as polymeric gene carrier. *Journal of Controlled Release*, 54: 39–48, 1998.

[6] Pouton, C.W., P. Lucas, B.J. Thomas, A.N. Uduehi, D.A. Milroy, and S.H. Moss. Polycation-DNA complexes for gene delivery: A comparison of the biopharmaceutical properties of cationic polypeptides and cationic lipids. *Journal of Controlled Release*, 53: 289–299, 1998.

[7] Wagner, E., C. Plank, K. Zatloukal, M. Cotten, and M.L. Birnstiel. Influenza virus hemagglutinin HA-2 N-terminal fusogenic peptides augment gene transfer by transferrin-polylysine-DNA complexes: Toward a synthetic virus-like gene-transfer vehicle. *Proceedings of the National Academy of Sciences of the United States of America*, 89: 7934–7938, 1992.

[8] Fajac, I., J.C. Allo, E. Souil, M. Merten, C. Pichon, C. Figarella, M. Monsigny, P. Briand, and P. Midoux. Histidylated polylysine as a synthetic vector for gene transfer into immortalized cystic fibrosis airway surface and airway gland serous cells. *Journal of Gene Medicine*, 2: 368–378, 2000.

[9] Fasman, G.D., M. Idelson, and E.R. Blout. The Synthesis and conformation of high molecular weight poly-ε-carbobenzyloxy-L-lysine and poly-L-lysine·HCl1,2. *Journal of the American Chemical Society*, 83: 709–712, 1961.

[10] Van Dijk-Wolthuis, W.N.E., L. van de Water, P. Van De Wetering, M.J. Van Steenbergen, J.J. Kettenes-Van Den Bosch, W.J.W. Schuyl and W.E. Hennink. Synthesis and characterization of poly-L-lysine with controlled low molecular weight. *Macromolecular Chemistry and Physics*, 198: 3893–3906, 1997.

[11] Inoue, Y., R. Kurihara, A. Tsuchida, M. Hasegawa, T. Nagashima, T. Mori, T. Niidome, Y. Katayama, and O. Okitsu. Efficient delivery of siRNA using dendritic poly(L-lysine) for loss-of-function analysis. *Journal of Controlled Release*, 126: 59–66, 2008.

[12] Watanabe, K., M. Harada-Shiba, A. Suzuki, R. Gokuden, R. Kurihara, Y. Sugao, T. Mori, Y. Katayama, and T. Niidome. In vivo siRNA delivery with dendritic poly(L-lysine) for the treatment of hypercholesterolemia. *Molecular BioSystems*, 5: 1306–1310, 2009.

[13] Matsumoto, S., R.J. Christie, N. Nishiyama, K. Miyata, A. Ishii, M. Oba, H. Koyama, Y. Yamasaki, and K. Kataoka. Environment-responsive block copolymer micelles with a disulfide crosslinked core for enhanced siRNA delivery. *Biomacromolecules*, 10: 119–127, 2009.

[14] Meyer, M., C. Dohmen, A. Philipp, D. Kiener, G. Maiwald, C. Scheu, M. Ogris, and E. Wagner. Synthesis and biological evaluation of a bioresponsive and endosomolytic siRNA-polymer conjugate. *Molecular Pharmaceutics*, 6: 752–762, 2009.

[15] Abbasi, M., H. Uludag, V. Incani, C.Y. Hsu, and A. Jeffery. Further investigation of lipid-substituted poly(L-Lysine) polymers for transfection of human skin fibroblasts. *Biomacromolecules*, 9: 1618–1630, 2008.

[16] Abbasi, M., A. Lavasanifar, L.G. Berthiaume, M. Weinfeld, and H. Uludag. Cationic polymer-mediated small interfering RNA delivery for P-glycoprotein down-regulation in tumor cells. *Cancer*, 116: 5544–5554, 2010.

[17] Shimizu, H., Y. Hori, S. Kaname, K. Yamada, N. Nishiyama, S. Matsumoto, K. Miyata, M. Oba, A. Yamada, K. Kataoka, and T. Fujita. siRNA-based therapy ameliorates glomerulonephritis. *Journal of the American Society of Nephrology*, 21: 622–633, 2010.

[18] Sato, A., S.W. Choi, M. Hirai, A. Yamayoshi, R. Moriyama, T. Yamano, M. Takagi, A. Kano, A. Shimamoto, and A. Maruyama. Polymer brush-stabilized polyplex for a siRNA carrier with long circulatory half-life. *Journal of Controlled Release*, 122: 209–216, 2007.

[19] Kano, A., K. Moriyama, T. Yamano, I. Nakamura, N. Shimada, and A. Maruyama. Grafting of poly(ethylene glycol) to poly-lysine augments its lifetime in blood circulation and accumulation in tumors without loss of the ability to associate with siRNA. *Journal of Controlled Release*, 149: 2–7, 2011.

[20] Singh, H.D., G. Wang, H. Uludag, and L.D. Unsworth. Poly-l-lysine-coated albumin nanoparticles: Stability, mechanism for increasing *in vitro* enzymatic resilience, and siRNA release characteristics. *Acta Biomaterialia*, 6: 4277–4284, 2010.

[21] Lee, S.K., M.S. Han, S. Asokan, and C.H. Tung. Effective gene silencing by multilayered siRNA-coated gold nanoparticles. *Small*, 7: 364–370, 2011.

10

Atelocollagen

10.1 Introduction

Atelocollagen was the first biomaterial with potential application as a gene delivery vector [1]. It is obtained by pepsin treatment from type I collagen of calf dermis [2, 3]. At the N- and C-terminals of the collagen molecules, there is an amino acid sequence called telopeptide which is responsible for collagen's antigenicity. However, atelocollagen obtained by pepsin treatment is free from telopeptides, is very low in immunogenicity and widely implicated for clinical purposes. At low temperatures, atelocollagen is in liquid form which favours complexation with nucleic acids. Since the surface of atelocollagen molecules is positively charged, the molecules can bond electrostatically with negatively charged nucleic acid molecules. When it is implanted in the body, it exhibits plasticity by which it initially becomes fibrous and later solidifies due to body temperature. At this time, gene vector and DNA are completely retained in the mesh structure of the matrix, and thereby protected from immunological reaction and enzymatic attack (Figure 10.1). Moreover, the atelocollagen solution containing DNA can be formed into beads, sponge, membrane, minipellet, etc. without heat processing or without using any organic solvent which is a major cause of deactivation of gene vector. The size of the complex particles can be manipulated by altering the ratio of nucleic acid to atelocollagen. When the concentration of atelocollagen is high, the complex persists locally for a long time, which is advantageous for a sustained release carrier. On the other hand, if the concentration of atelocollagen is low, the diameter of the complex particles is 100–300 nm, which is considered adequate for systemic treatment.

S. Nimesh and R. Chandra,
Theory, Techniques and Applications of Nanotechnology in Gene Silencing, 119–128.

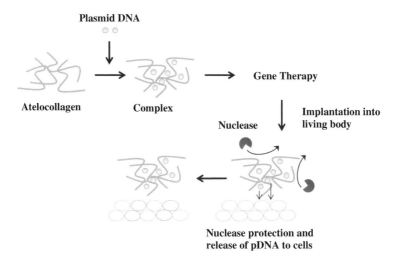

Plasmid DNA

Atelocollagen   **Complex**   **Gene Therapy**

Nuclease   **Implantation into living body**

Nuclease protection and release of pDNA to cells

Fig. 10.1 Mechanism of atelocollagen-mediated gene delivery.

Moreover, the siRNA-atelocollagen complexes can also be pre-coated on a micro-well plate into which the cells are seeded [4, 5]. This method highly facilitates cellular uptake of siRNA-atelocollagen complex thereby enhancing gene silencing. One of the major hurdles for systemic treatment using siRNA is its rapid degradation by endonucleases. However, siRNA complexed with atelocollagen is resistant to nucleases and is transduced efficiently into cells, thereby allowing long-term gene silencing [5].

10.2   Atelocollagen-Mediated siRNA Delivery

The potential of atelocollagen-mediated siRNA transfer have been tested successfully for gene silencing *in vivo*, using nude mice bearing luciferase producing tumor cells [4]. Non-invasive *in vivo* bioluminescence imaging analysis can be utilized to evaluate the efficiency of delivery of siRNA against luciferase mRNA (luc-siRNA) into tumor cells by suppression of luciferase expression and production of photons from tumor cell inoculated mice. With this strategy, subcutaneous injection of the luc-siRNA-atelocollagen complex showed a sustained inhibition of luciferase expression from tumor cells xenografted back into mice [4]. In the case of inhibition studies of tumor growth, intratumoral injection of a HST-1/FGF-4-siRNA-atelocollagen complex presented efficient inhibition of tumor growth in an orthotopic xenograft model of a human testicular cancer. Takei *et al.* showed that radiolabeled siRNA mixed with atelocollagen existed in the tumors for at least one week and remained intact [6]. Treatment with VEGF-siRNA-atelocollagen complex dramatically suppressed tumor angiogenesis and tumor growth in a PC-3 subcutaneously xenograft model.

Systemic delivery of siRNA can also be achieved by an atelocollagen complex delivered by intravenous injection. In recent reports, in order to estimate the effectiveness of systemic delivery of siRNA, a mouse model of bone metastatic human prostate cancer was prepared [7]. In this model, bone metastases developed in the jaws and/or legs of the mice were detected by non-invasive *in vivo* bioluminescence imaging analysis. When administered with the luc-siRNA-atelocollagen complex in the mice, bioluminescence at day 1 post-treatment was inhibited by 80–90% in the whole body, including the bone metastases, when compared with before treatment [7]. On the other hand, the bioluminescent signals from the mice treated with atelocollagen alone increased, and those treated with luciferase siRNA alone either had no effect or slightly suppressed luciferase expression. Furthermore, in order to assess the inhibition of tumor growth on bone metastasis by the atelocollagen-mediated siRNA delivery system, human enhancer of zeste homolog 2 (EZH2) and human phosphoinositide 3′-hydroxykinase p110α subunit (p110α) siRNA-atelocollagen complexes were administered intravenously into mice on days three, six and nine post-injection of luciferase-producing human prostate cancer cells [7]. As a result, there was significant inhibition of tumor growth in bone tissues in EZH2 and p110α siRNA-atelocollagen treated groups at experimental day 28. In addition, upregulation of serum interleukin-12 and interferon-α level was not associated with systemic injection of the siRNA-atelocollagen complex. Thus, for treatment of bone metastasis of prostate cancer, a new atelocollagen-mediated systemic delivery method could be a reliable and safe approach to the achievement of maximal function of siRNA *in vivo*.

To determine whether blockade of PAR-2 by RNAi influences pancreatic tumor growth, PAR-2 siRNA was delivered with atelocollagen [8]. Initially, siRNAs-targeting human PAR-2 was constructed, followed by cell proliferation assays of Panc1 human pancreatic cancer cell line with these siRNAs. Cells treated with PAR-2 siRNA no. 3 and SLIGKV peptide showed significantly lower cell proliferation than the controls suggesting that downregulation of PAR-2 by siRNA suppressed Panc1 cell proliferation *in vitro*. Intratumoral treatment with these PAR-2 siRNAs-atelocollagen complexes was also performed in a xenograft model with nude mice and Panc1 cells. siRNAs against human PAR-2 inhibited the proliferation of Panc1 cells, whereas control scramble siRNAs had no effect on proliferation [9]. The PAR-2 siRNAs dramatically suppressed tumor growth in the xenograft model. PAR-2 specific siRNA inhibited growth of human pancreatic cancer cells both *in vitro* and *in vivo* suggesting that blockade of PAR-2 signaling by siRNA may be a novel strategy to treat pancreatic cancer. Kinouchi *et al.* reported the effectiveness of *in vivo* siRNA delivery into skeletal muscles of normal or diseased mice through nanoparticle formation of chemically unmodified siRNAs with atelocollagen [10]. Atelocollagen-mediated local application of GDF8 siRNA26-targeting myostatin, a negative regulator of skeletal muscle growth, in mouse skeletal muscles or intravenously, caused a marked increase in the muscle mass within a few weeks after application. These results imply that atelocollagen-mediated application of siRNAs is a powerful tool for future therapeutic use for diseases including muscular atrophy.

With the objective to apply siRNA technology for therapy of hrHPV related cancers, siRNA sequences targeting mRNA coding HPV16 E6 and E7 that have high levels of RNAi activity and minimal off-target effects were delivered with atelocollagen [11]. SiHa cells were subcutaneously inoculated into NOD/SCID mice. Six weeks after inoculation, palpable tumors formed in all the mice, and control or 752 siRNA complexed with atelocollagen was directly injected into each tumor every seven days. Tumors treated with siRNA 752 had significantly lower growth than those treated with control siRNA. Hence, E6 and E7 siRNAs complexed with atelocollagen are promising therapeutic agents for treatment of virus-related cancer.

Heparan sulfate proteoglycan syndecan-1 (CD138) is well known to be associated with cell proliferation, adhesion and migration in various types of malignancies. Shimada *et al.* investigated the role of syndecan-1 in human prostate cancer [12]. Syndecan-1 expression was much higher in the androgen independent prostate cancer cell lines DU145 and PC3, rather than the androgen-dependent LNCaP, but the level in LNCaP was upregulated in response to long-term culture under androgen deprivation. Nude mice with subcutaneous inoculation of PC3 cells receiving a localized injection of syndecan-1 siRNA mixed with atelocollagen showed syndecan-1 downregulation, producing approximately two-fold decrease in tumor size and vessel density. Knockdown effects by single injection of syndecan-1 siRNA-atelocollagen on subcutaneously inoculated prostate cancer cells could be sustained for approximately 5–7 days. Moreover, single injection of syndecan-1 siRNA-atelocollagen strongly downregulated syndecan-1, NOX2 and VEGF protein expression resulting in apoptosis induction and tumor volume reduction.

Mu *et al.* showed a systemic delivery method of siRNA specific to pre-grown solid tumors via atelocollagen [13]. Atelocollagen facilitated the selective uptake of siRNA into the tumors when a siRNA-atelocollagen complex was administered intravenously to mice. A Bcl-xL protein was chosen as a model target to prove the therapeutic efficacy of the atelocollagen-mediated method. Bcl-xL acts as an antiapoptotic factor, which is overexpressed in many cancers, including prostate cancer. One of the four designed siRNAs to human Bcl-xL potently inhibited the expression of Bcl-xL by the PC-3 human prostate cancer cell line *in vitro*, leading to cell apoptosis. Intravenous injections administered for three consecutive days (siRNA 100 μg/injection per day with atelocollagen) effectively downregulated Bcl-xL expression in PC-3 xenografts. Four series of three consecutive days of intravenous injections each administered for a total of 12 injections, significantly inhibited tumor growth when the treatment was combined with cisplatin (2 mg/kg). All the tumors treated with Bcl-xL siRNA-atelocollagen complex via both intravenous and intra-tumoral injections showed terminal deoxynucleotidyl transferase-mediated dUTP nickend labeling positive apoptosis. There were no severe side effects such as interferon-α induction and liver or renal damage in mice. The results indicate that systemic delivery of siRNA via atelocollagen, which specifically targets tumors, is safe and feasible for cancer therapy.

Several factors contribute to the development of prostate cancer including somatic mutations of the androgen receptor (AR) or AR amplification. The effect of synthetic siRNA targeting AR (siAR) has been investigated on the *in vitro* and *in vivo* growth of human androgen-independent prostate cancer (AIPC) cells expressing mutated AR [14]. Administration of siAR-atelocollagen complexes into mouse tail veins every three days for a total of five injections resulted in significant reduction in the size of subcutaneously xenografted 22Rv1 tumors, compared with control groups. Additionally, the expression of the AR in excised tumor tissues was notably suppressed in siAR-atelocollagen complex administration groups compared with control groups. The study suggests that AIPC is likely to have an activated AR signal transduction system, and that treatment with AR-targeted siRNA-atelocollagen is an effective strategy for the treatment of AIPC. In another study, gene FABP5 which encodes cutaneous fatty acid binding protein (C-FABP) that is upregulated in prostate cancer was selected as a target to silence with FABP5 siRNA-atelocollagen complexes [15]. The complexes were injected around tumor masses produced by PC-3M cells in Balb/c nude mice and compared with the effect of non-specific scrambled siRNA in atelo-collagen. At autopsy, the average size of tumors from the groups treated with 10 and 15 μM siRNA in atelocollagen was significantly ($p = 0.02$) reduced by more than 3-fold, when compared to the controls. Immunohistochemistry and western blotting revealed that the levels of C-FABP expression in tumors from mice treated with 10 and 15 μM dosages were lower than those from the other groups. These data demonstrate that FABP5 siRNA delivered by ate-locollagen to the external environment surrounding a tumor mass can effec-tively inhibit prostate cancer cell growth in nude mice when administered in a dose-dependent manner at concentrations of > 10 μM. Recent studies indicate that Akt (serine/threonine kinase) which exist in three isoforms: Akt1, Akt2, and Akt3 is often constitutively active in many types of human carcinomas including prostate cancer. Synthetic siRNA against each Akt on transfection with atelocollagen into PC-3 cells expressing all Akt isoforms revealed reduc-tion in the expression of the corresponding Akt isoform by 36–63% [16]. Furthermore, administration of siRNA-atelocollagen complexes into mouse tail veins every three days for a total of five injections significantly reduced the size of subcutaneously xenografted PC-3 tumors, compared with control groups. Moreover, the expression of the corresponding Akt isoform in excised tumor tissues was notably suppressed in the groups administered with siRNA-atelocollagen complex compared with the control groups.

A novel human AlkB homologue, ALKBH3, which contributes to prostate cancer development, was investigated as possible target in treatment of lung cancer [17]. The effects of ALKBH3 gene silencing *in vivo* was investigated in an animal model of intraperitoneal inoculation of A549 and RERF-LC-AI cells using nude mice. The numbers of tumors formed in the peritoneum of nude mouse model were significantly decreased in mice injected with ALKBH3 siRNA-atelocollagen complexes as compared with mice receiving control siRNA. The study concluded that ALKBH3 contributes significantly to cancer cell survival and may be a therapeutic target for human adenocarcinoma of the lung. The EWS/Fli-1 fusion gene, a product of the translocation t (11;22, q24;q12), is detected in 85% of Ewing sarcomas and primitive neuroectodermal tumors. Recently, a study was designed to develop a novel EWS/Fli-1 blockade system using RNAi and test its application for inhibiting the proliferation of Ewing sarcoma cells *in vitro* and the treatment of mouse tumor xenografts *in vivo* [18]. siRNA targeting the breakpoint of EWS/Fli-1 was prepared by modification with an aromatic compound at the 3′-end. It was observed that the siRNA targeting EWS/Fli-1 significantly suppressed the expression of EWS/Fli-1 protein sequence specifically and also reduced the expression of c-Myc protein in Ewing sarcoma cells. Furthermore, the inhibition of EWS/Fli-1 expression efficiently inhibited the proliferation of the transfected cells but did not induce apoptotic cell death. Administration of the siRNA-atelocollagen significantly inhibited the tumor growth of TC-135, a Ewing sarcoma cell line, which was subcutaneously xenografted into mice. Moreover, modification of the 30-end with an aromatic compound improved its efficiency *in vivo*. This study proposed specific downregulation of EWS/Fli-1 by RNAi as a possible approach for the treatment of Ewing sarcoma.

10.3 Future Perspectives

The systemic delivery of siRNA utilizing a tissue-specific or cell-specific gene promoter vector or specific antibody-conjugated carriers, thus reducing applied dose of siRNA and resulting in decreased side effects, will need to be developed. For designing target specific vectors, angiogenesis and metastasis can be exploited for the differences between cancerous cells and normal cells, which include uncontrolled proliferation, insensitivity to negative growth regulation and anti-growth signals. A growing list of unique cancer markers is underway which forms the basis of key interactions between the siRNA-carrier complexes and cancer cells. Although, atelocollagen has been implicated in numerous anti-tumour studies in xenograft models, it is noteworthy that human xenograft models in immunodeficient mice frequently overpredict activity and underpredict toxicity. To further address such issues related to safety and efficacy of delivered siRNA, extensive and careful research will need to be undertaken.

10.4 Conclusions

Atelocollagen possesses minimal antigenicity and high transfection efficiency and has been extensively investigated for *in vivo* siRNA delivery to variety of tumor xenograft models. Exciting results leading to suppression of tumor growth via siRNA-atelocollagen complexes proposes promising future for further development. However, lack of insufficient data regarding the mechanism and off-target effects pose major obstacles towards development of siRNA-based therapeutics. Hence, exhaustive details pertaining to efficacy and pharmacokinetics of siRNA-atelocollagen systems will pave the way towards the successful implication of clinical applications.

References

[1] Ochiya, T., Y. Takahama, S. Nagahara, Y. Sumita, A. Hisada, H. Itoh, Y. Nagai, and M. Terada. New delivery system for plasmid DNA *in vivo* using atelocollagen as a carrier material: The minipellet. *Nature Medicine*, 5: 707–710, 1999.

[2] Ochiya, T., S. Nagahara, A. Sano, H. Itoh, and M. Terada. Biomaterials for gene delivery: Atelocollagen-mediated controlled release of molecular medicines. *Current Gene Therapy*, 1: 31–52, 2001.

[3] Sano, A., M. Maeda, S. Nagahara, T. Ochiya, K. Honma, H. Itoh, T. Miyata, and K. Fujioka. Atelocollagen for protein and gene delivery. *Advanced Drug Delivery Reviews*, 55: 1651–1677, 2003.

[4] Minakuchi, Y., F. Takeshita, N. Kosaka, H. Sasaki, Y. Yamamoto, M. Kouno, K. Honma, S. Nagahara, K. Hanai, A. Sano, T. Kato, M. Terada, and T. Ochiya. Atelocollagen-mediated synthetic small interfering RNA delivery for effective gene silencing *in vitro* and *in vivo*. *Nucleic Acids Research*, 32: e109, 2004.

[5] Honma, K., T. Miyata, and T. Ochiya. The role of atelocollagen-based cell transfection array in high-throughput screening of gene functions and in drug discovery. *Current Drug Discovery Technologies*, 1: 287–294, 2004.

[6] Takei, Y., K. Kadomatsu, Y. Yuzawa, S. Matsuo, and T. Muramatsu. A small interfering RNA targeting vascular endothelial growth factor as cancer therapeutics. *Cancer Research*, 64: 3365–3370, 2004.

[7] Takeshita, F., Y. Minakuchi, S. Nagahara, K. Honma, H. Sasaki, K. Hirai, T. Teratani, N. Namatame, Y. Yamamoto, K. Hanai, T. Kato, A. Sano, and T. Ochiya. Efficient delivery of small interfering RNA to bone-metastatic tumors by using atelocollagen *in vivo*. *Proceedings of the National Academy of Sciences of the United States of America*, 102: 12177–12182, 2005.

[8] Iwaki, K., K. Shibata, M. Ohta, Y. Endo, H. Uchida, M. Tominaga, R. Okunaga, S. Kai, and S. Kitano. A small interfering RNA targeting proteinase-activated receptor-2 is effective in suppression of tumor growth in a panc1 xenograft model. *International Journal of Cancer*, 122: 658–663, 2008.

[9] Tamai, T., M. Watanabe, Y. Hatanaka, H. Tsujiwaki, N. Nishioka, and K. Matsukawa. Formation of metal nanoparticles on the surface of polymer particles incorporating polysilane by UV irradiation. *Langmuir*, 24: 14203–14208, 2008.

[10] Kinouchi, N., Y. Ohsawa, N. Ishimaru, H. Ohuchi, Y. Sunada, Y. Hayashi, Y. Tanimoto, K. Moriyama, and S. Noji. Atelocollagen-mediated local and systemic applications of myostatin-targeting siRNA increase skeletal muscle mass. *Gene Therapy*, 15: 1126–1130, 2008.

[11] Yamato, K., T. Yamada, M. Kizaki, K. Ui-Tei, Y. Natori, M. Fujino, T. Nishihara, Y. Ikeda, Y. Nasu, K. Saigo, and M. Yoshinouchi. New highly potent and specific E6 and E7 siRNAs for treatment of HPV16 positive cervical cancer. *Cancer Gene Therapy*, 15: 140–153, 2008.

[12] Shimada, K., M. Nakamura, M.A. De Velasco, M. Tanaka, Y. Ouji, and N. Konishi. Syndecan-1, a new target molecule involved in progression of androgen-independent prostate cancer. *Cancer Science*, 100: 1248–1254, 2009.

[13] Mu, P., S. Nagahara, N. Makita, Y. Tarumi, K. Kadomatsu, and Y. Takei. Systemic delivery of siRNA specific to tumor mediated by atelocollagen: Combined therapy using siRNA targeting Bcl-xL and cisplatin against prostate cancer. *International Journal of Cancer*, 125: 2978–2990, 2009.

[14] Azuma, K., K. Nakashiro, T. Sasaki, H. Goda, J. Onodera, N. Tanji, M. Yokoyama, and H. Hamakawa. Anti-tumor effect of small interfering RNA targeting the androgen receptor in human androgen-independent prostate cancer cells. *Biochemical and Biophysical Research Communications*, 391: 1075–1079, 2010.

[15] Forootan, S.S., Z.Z. Bao, F.S. Forootan, L. Kamalian, Y. Zhang, A. Bee, C.S. Foster, and Y. Ke. Atelocollagen-delivered siRNA targeting the FABP5 gene as an experimental therapy for prostate cancer in mouse xenografts. *International Journal of Oncology*, 36: 69–76, 2010.

[16] Sasaki, T., K. Nakashiro, H. Tanaka, K. Azuma, H. Goda, S. Hara, J. Onodera, I. Fujimoto, N. Tanji, M. Yokoyama, and H. Hamakawa. Knockdown of Akt isoforms by RNA silencing suppresses the growth of human prostate cancer cells in vitro and in vivo. *Biochemical and Biophysical Research Communications*, 399: 79–83, 2010.

[17] Tasaki, M., K. Shimada, H. Kimura, K. Tsujikawa, and N. Konishi. ALKBH3, a human AlkB homologue, contributes to cell survival in human non-small-cell lung cancer. *British Journal of Cancer*, 104: 700–706, 2011.

[18] Takigami, I., T. Ohno, Y. Kitade, A. Hara, A. Nagano, G. Kawai, M. Saitou, A. Matsuhashi, K. Yamada, and K. Shimizu. Synthetic siRNA targeting the breakpoint of EWS/Fli-1 inhibits growth of Ewing sarcoma xenografts in a mouse model. *International Journal of Cancer*, 128: 216–226, 2011.

11

Protamine

11.1 Introduction

Protamines are small, arginine-rich, nuclear proteins that replace histones in the late haploid phase of spermatogenesis. On an average, 67% of the amino acid composition in protamine is arginine. It is a naturally occurring substance found only in the sperm and is purified from the mature testes of fish, usually salmon with an average MW of 4500 Da. Protamine's role in sperm is to bind DNA, assist in the formation of a compact structure followed by delivery of DNA to the nucleus of the egg after fertilization. This unique functionality overcomes a major obstacle in gene therapy by non-viral vectors i.e. the efficient delivery of DNA from the cytoplasm into the nucleus. It is routinely used to reverse the anticoagulant action of heparin after cardiac or vascular surgeries. The polycationic protamine combines with the polyanionic heparin through an electrostatic interaction, thereby neutralizing the anticoagulant functions of heparin. Protamine sulfate have been shown to be a safer and more appropriate alternative to poly-L-lysine for condensation, as well as the delivery of plasmid DNA to the nucleus. Sorgi *et al.* described the ability of several different protamine species in enhancing the transfection activity of cationic lipids [1]. The structural differences between the protamines account for the differences among various salt forms of protamine. The appearance of lysine residues within the protamine molecule correlated well with a reduction in the binding affinity to plasmid DNA, as well as an observed loss in the transfection activity. In another study, You *et al.* observed improvement in transfection efficiency by adding protamine to plasmid DNA solution before

S. Nimesh and R. Chandra,
Theory, Techniques and Applications of Nanotechnology in Gene Silencing, 129–138.

the formation of DNA-lipid vesicle complexes [2]. Both free-base protamine and protamine sulfate revealed better transfection efficiency and expression level, where the optimal amount of the two protamines was different. The increment in the transfection efficiency and gene expression level was at most 20-fold compared with those using only dimethyldioctadecyl ammonium bromide lipid vesicles. Protamines were thought to protect DNA from degradation by DNAse and promote delivery into the nucleus.

11.2 Protamine-siRNA Complexes for Gene Silencing

Protamine has found numerous applications such as in compact liposomes-DNA complex (LPD); where liposomes interact with DNA condensed by protamine resulting in lipid rearrangement and the formation of LPD. Use of protamine for siRNA delivery was first reported by Song *et al.* [3]. Protamine was fused to F105, the heavy chain of a Fab fragment specific for the HIV envelope protein gp160, and was used to deliver siRNAs to silence gene expression specifically in HIV-infected cells or cells transfected to express HIV envelope glycoprotein gp160 (HIV env) in COS cells. Uptake of the Fab-protamine-complexed siRNA by Jurkat cells depended on HIV infection (and therefore on gp160 expression) along with the presence of F105 and protamine in the siRNA complex. Further, siRNA-mediated silencing of GFP expression in HeLa cells also depended on HIV infection and siRNA delivery by a functional F105-protamine-siRNA complex [3]. Additionally, this construct was used to silence expression of the major HIV core protein p24 in cultured HIV-1-infected T cells, which are usually difficult to transfect with nucleic acids. siRNAs delivered using the F105-protamine fusion targeting mRNA for the oncogenes c-myc, MDM2 and VEGF efficiently reduced proliferation in the gp160-expressing B16 melanoma cells. No stimulation of immune response was observed upon monitoring the expression analysis of interferon-β and two interferon responsive genes, $20'-50'$ oligoadenylate synthetase (OAS1) and signal transducer, activator of transcription 1(Stat1). Further, the delivery efficacy of Fab-protamine-fusion was demonstrated by using another antibody. A protamine fragment was fused to a single-chain antibody specific for ErbB2 and the construct was expressed in insect cells. Delivery of siRNA with the fusion protein targeted to Ku70 (a protein involved in DNA repair) efficiently silenced the target in ErbB2-expressing breast cancer cells. Fab-protamine-fusion was further assessed for *in vivo* delivery efficacy to transport fluorescent siRNA into gp160-B16 cells subcutaneously implanted into the right flanks of syngeneic C57/BL6 mice [3]. siRNA was not detected in the mice when delivered in PBS, however, oligofectamine complexed siRNA was detected in tumor cells and in adjacent tissue. On the contrary, F105-protamine specifically delivered FITC-siRNA to gp160$^+$ tumor cells, but not to adjacent tissue. Furthermore, intratumoral injection of a mixture of siRNAs against c-myc, MDM2 and VEGF genes in mice B16 or gp160-B16 tumor cells significantly reduced the tumor growth when compared to those treated with siRNAs alone. Intratumoral injection of siRNA was slightly more efficient than systemic intravenous delivery.

Lymphocytes and other primary blood cells are resistant to transfection using conventional transfection reagents (e.g., cationic lipids and polymers) both in *in vitro* and *in vivo*. Peer *et al.* showed that antibody-protamine fusion proteins targeting the human integrin lymphocyte function-associated antigen-1 (LFA-1) efficiently deliver siRNAs and specifically induce silencing in primary lymphocytes, monocytes, and dendritic cells [4]. An AL-57 scFv-protamine fragment fusion protein (AL-57-PF) was designed that preferentially recognizes activation dependent high affinity conformation of LFA-1. Flow cytometry analysis revealed that AL-57-PF potently delivered Cy3 labeled siRNA to all subsets of stimulated peripheral blood mononuclear cells (PBMC). Further, AL-57-PF was used to deliver siRNAs to induce silencing of the ubiquitously expressed Ku70 gene selectively to high affinity LFA-1-expressing cells. Ku70-siRNA delivered by AL-57-PF induced silencing specifically in stimulated cells which was easily detectable with 100 pmol of siRNA and plateaued at ~2,000 pmol. Furthermore, to achieve *in vitro* selective attenuation of CCR5 expression in activated lymphocytes which is upregulated in rheumatoid arthritis and transplant rejection, CCR5-siRNA was delivered by LFA-1 antibody fusion protein. As expected, CCR5 was significantly reduced by CCR5-siRNA delivered by AL-57-PF only in stimulated lymphocytes. These results demonstrate potent and selective gene silencing only in activated PBMC and T lymphocytes after siRNA delivery with AL-57-PF that are normally resistant to transfection. Moreover, the siRNA-fusion protein complexes do not cause lymphocyte activation or induce IFN responses. K562 cells expressing latent WT or constitutively activated LFA-1 engrafted in the lungs of SCID mice are selectively targeted by intravenously injected fusion protein-siRNA complexes, demonstrating the potential *in vivo* applicability of LFA-1-directed siRNA delivery.

In a study, self-assembled nanoparticles were developed that efficiently deliver siRNA to the tumor by intravenous administration [5]. The nanoparticles were obtained by mixing calf thymus DNA, siRNA, protamine, and lipids, followed by post-modification with PEG, ligand and anisamide. Four hours after intravenous injection of the formulation into a NCI-H460 xenograft model, 70–80% of injected siRNA/g accumulated in the tumor, ~10% was detected in the liver and ~20% recovered in the lung. Confocal microscopy revealed that fluorescently labeled siRNA was efficiently delivered into the cytoplasm of the sigma receptor expressing NCI-H460 xenograft tumor by the targeted nanoparticles, whereas free siRNA and non-targeted nanoparticles showed little uptake. Three consecutive daily doses (1.2 mg/kg) of siRNA complexed in the targeted nanoparticles silenced the EGFR in the tumor and induced ~15% tumor cell apoptosis. Up to 40% tumor growth inhibition was achieved by treatment with targeted nanoparticles, while complete inhibition lasted for one week when combined with cisplatin [5]. The serum level of liver enzymes and body weight monitored during the treatment indicated a low level of toxicity of the formulation while the carrier itself showed little immunotoxicity. The nanoparticle formulation provided the advantages of an almost complete encapsulation efficiency (>95%), the highest tumor delivery efficiency so far reported (70–80% ID/g at 4 hours) and relatively easy preparation. Further, to reduce the immunotoxicity of the formulation, the same group investigated the probability of using hyaluronic acid to improve the nanoparticle formation [6]. A nanoparticle formulation [liposomes-protamine-hyaluronic acid nanoparticle (LPH-NP)] was developed for systemically delivering siRNA into the tumor. The LPH-NP was prepared in a self-assembling process by mixing protamine with a mixture of siRNA and hyaluronic acid to obtain a negatively charged complex, followed by coating with cationic lipids via charge-charge interaction to yield the LPH-NP. The LPH-NP was further modified by DSPE-PEG or DSPE-PEG-anisamide by the post-insertion method. Anisamide is a well known targeting ligand for the sigma receptor overexpressed in the B16F10 melanoma cells. The physicochemical studies suggested of particle size, zeta potential and siRNA encapsulation efficiency of the formulation to be approximately 115 nm, +25 mV and 90%, respectively.

B16F10 cells, stably transducing a luciferase gene, were used to evaluate the luciferase gene silencing activity [6]. The targeted LPH-NP showed more than 60% silencing of luciferase gene expression. Upon systemic delivery, a single intravenous injection (0.15 mg siRNA/kg) of targeted LPH-NP (PEGylated with ligand) silenced 80% of luciferase activity in the metastatic B16F10 tumor in the lung [6]. Additionally, the targeted LPH-NP also showed very little immunotoxicity in a wide dose range (0.15–1.2 mg siRNA/kg). The targeted LPH-NP showed similar characteristics and gene silencing activity compared to the targeted LPD-NP, while significantly improving the therapeutic window by at least 2.7-fold.

In cancer therapy, protein-based drug delivery to intracellular targets is largely dictated by the specific binding to a tumor associated antigen abundantly expressed on the cell surface, followed by quantitative internalization through receptor- mediated endocytosis. The overall efficacy is outcome of several parameters including degree of tumor localization, the binding to the tumor associated antigen, its endocytosis, and finally the transfer of the payload from the endosome to the cytoplasm. The epithelial cell adhesion molecule (EpCAM) is a 40 kDa transmembrane glycoprotein upregulated in solid tumors, while minimally expressed in normal epithelia [7]. Upon interaction with antibodies, EpCAM is rapidly internalized and thus ideally suited for delivery of anticancer agents to intracellular targets [8]. Ankyrin repeat protein (DARPins) is a new class of binding molecules with excellent biochemical properties, which can be ideally employed for tumor targeting [9]. Recently, Winkler *et al.* reported the use of EpCAM-specific DARPin (C9) to generate fusion proteins with protamine-1 for complexation of siRNA complementary to anti-apoptotic bcl-2, a potent inhibitor of apoptosis implicated in cancer drug resistance [10]. The fusion proteins were capable of complexing four to five siRNA molecules per protamine, and retained full binding specificity for EpCAM as shown in the MCF-7 breast carcinoma cells. The confocal microscopy suggested of a punctuate fluorescence distribution pattern indicative of endo/lysosomal localization of the siRNA, alongwith diffuse fluorescence distributed in the cytoplasm. Further, the bcl-2 mRNA was significantly reduced in a dose-dependent manner upon treatment with siRNA complexed to protamine fusion protein compared to a scrambled control siRNA sequence. Delivery of siRNA with the DARPin nanocomplexes also led to bcl-2 mRNA and protein downregulation in EpCAM-positive HT-29 colon carcinoma cells. Inhibition of bcl-2 expression facilitated tumor cell apoptosis as shown by increased sensitivity to the anticancer agent doxorubicin [10].

Low molecular weight protamine (LMWP) possesses cell translocating capability and can efficiently deliver large protein molecules into tumors, thereby suppressing tumor growth *in vivo* [11]. LMWP is synthesized by enzymatic digestion of native protamine with thermolysin. Choi *et al.* demonstrated the potential application of LMWP to achieve siRNA transduction into cells with an efficiency equivalent to that of TAT_{47-57} peptide, known as one of potent CPPs [12]. Tracking studies suggested that fluorescently tagged siRNAs were localized with the peptide in the cytoplasm shortly after incubation of LMWP-siRNA complex with carcinoma cells. The enhanced cellular internalization of siRNA achieved using the LMWP resulted in significant downregulation of model protein luciferase as well as therapeutic cancer target, VEGF expression. *In vivo* studies with tumor-bearing mice further demonstrated that the peptide could carry and localize siRNA inside tumors and inhibit the expression of VEGF through systemic application of the peptide complex, thereby suppressing tumor growth [12]. Additionally, no increase in the serum level of inflammatory cytokines including interferon-α and interleukin-12 was observed on treatment with LMWP-siRNA complex, suggesting of nonimmunogenicity. The authors proposed the LMWP-based systemic delivery method to be a reliable and safe approach to maximize effectiveness of therapeutic siRNA for treatment of cancer and other diseases.

Podocytes are specialized terminally differentiated epithelial cells comprising of the glomerular filtration barrier of the kidney; injured in several glomerular diseases. To manipulate the *in vivo* gene expression in podocytes, Hauser *et al.* exploited their active endocytotic machinery and developed a method for targeted delivery of siRNA [13]. Antimouse podocyte antibody was generated that binds to rat and mouse podocytes *in vivo*. The polyclonal IgG antibody was cleaved into monovalent fragments, while preserving the antigen recognition sites. One Neutravidin molecule was linked to each monovalent IgG via the available sulfohydryl group, followed by linkage of protamine via biotin. The delivery system was named *shamporter* (*sh*eep *a*nti*m*ouse *po*docyte transpo*rter*). Tail vein administration of *shamporter* coupled with either nephrin siRNA or TRPC6 siRNA into normal rats significantly reduced the protein levels of nephrin or TRPC6 respectively, assayed by western blot analysis and immunostaining [13]. The effect was highly target-specific as other podocyte-specific genes remained unaltered. Combination of *shamporter* with nephrin siRNA induced transient proteinuria in rats. In contrast, control rats injected with shamporter coupled to control-siRNA showed no changes. These results show for the first time that siRNA can be delivered efficiently and specifically to podocytes *in vivo* using an antibody delivery system.

11.3 Future Perspectives

Numerous studies demonstrated that cell type-specific antibodies can be used to systemically deliver siRNA *in vivo*. The demonstrated specificity, efficiency of gene silencing and the lack of non-specific immune activation or systemic toxicity are encouraging. Further emphasis is needed to determine how silencing is affected by the internalizing activity of the targeting antibody and by the release of the complex in the cytosol. *In vivo* biodistribution studies are also needed to further determine target specificity and penetrance of these delivery systems. Finally, the efficacy and specificity of *in vivo* siRNA delivery depends on the targeting antibodies or other cell type-specific affinity ligands. Overall, the studies citing the application of protamine-based fusion complexes demonstrated that the barriers for *in vivo* siRNA delivery can be overcome, significantly raising the likelihood of siRNA-based therapeutics against a broad spectrum of human diseases.

11.4 Conclusion

These studies provide sufficient evidence about the potential for *in vivo* systemic, cell type-specific, antibody-mediated siRNA delivery. The protein fusion strategy is simple and broadly applicable to antibodies and antibody fragments. Moreover, the size of siRNA-protein fusion complexes is above the threshold for renal clearance, and thus they are expected to have longer circulating half-life than the naked siRNA. Further studies are needed to determine pharmacokinetics and biodistribution (including penetrance and relative accumulation in target cells/tissues) of the fusion complex. On an average, protamine protein binds to a limited number (<6) of siRNA molecules, and only a fraction of the surface-bound scFvs are likely to be internalized and released into the cytoplasm. The actual number of siRNA entering the RNAi pathway may be heavily dependent on receptor density, antibody affinity, and the rate of endocytosis for a given antibody-receptor pair. Small nucleic acid-binding proteins other than protamine may also be studied to increase the overall production and applications of the fusion protein and to further optimize siRNA delivery systems.

References

[1] Sorgi, F.L., S. Bhattacharya, and L. Huang. Protamine sulfate enhances lipid-mediated gene transfer. *Gene Therapy*, 4: 961–968, 1997.

[2] You, J., M. Kamihira, and S. Iijima. Enhancement of transfection efficiency by protamine in DDAB lipid vesicle-mediated gene transfer. *Journal of Biochemistry*, 125: 1160–1167, 1999.

[3] Song, E., P. Zhu, S.K. Lee, D. Chowdhury, S. Kussman, D.M. Dykxhoorn, Y. Feng, D. Palliser, D.B. Weiner, P. Shankar, W.A. Marasco, and J. Lieberman. Antibody mediated *in vivo* delivery of small interfering RNAs via cell-surface receptors. *Nature Biotechnology*, 23: 709–717, 2005.

[4] Peer, D., P. Zhu, C.V. Carman, J. Lieberman, and M. Shimaoka. Selective gene silencing in activated leukocytes by targeting siRNAs to the integrin lymphocyte function-associated antigen-1. *Proceedings of the National Academy of Sciences of the United States of America*, 104: 4095–4100, 2007.

[5] Li, S.D., Y.C. Chen, M.J. Hackett, and L. Huang. Tumor-targeted delivery of siRNA by self-assembled nanoparticles. *Molecular Therapy*, 16: 163–169, 2008.

[6] Chono, S., S.D. Li, C.C. Conwell, and L. Huang. An efficient and low immunostimulatory nanoparticle formulation for systemic siRNA delivery to the tumor. *Journal of Controlled Release*, 131: 64–69, 2008.

[7] Went, P., M. Vasei, L. Bubendorf, L. Terracciano, L. Tornillo, U. Riede, J. Kononen, R. Simon, G. Sauter. and P.A. Baeuerle. Frequent high-level expression of the immunotherapeutic target Ep-CAM in colon, stomach, prostate and lung cancers. *British Journal of Cancer*, 94: 128–135, 2006.

[8] Di Paolo, C., J. Willuda, S. Kubetzko, I. Lauffer, D. Tschudi, R. Waibel, A. Pluckthun, R.A. Stahel, and U. Zangemeister-Wittke. A recombinant immunotoxin derived from a humanized epithelial cell adhesion molecule-specific single-chain antibody fragment has potent and selective antitumor activity. *Clinical Cancer Research*, 9: 2837–2848, 2003.

[9] Binz, H.K., P. Amstutz, A. Kohl, M.T. Stumpp, C. Briand, P. Forrer, M.G. Grutter, and A. Pluckthun. High-affinity binders selected from designed ankyrin repeat protein libraries. *Nature Biotechnology*, 22: 575–582, 2004.

[10] Winkler, J., P. Martin-Killias, A. Pluckthun, and U. Zangemeister-Wittke. EpCAM-targeted delivery of nanocomplexed siRNA to tumor cells with designed ankyrin repeat proteins. *Molecular Cancer Therapeutics*, 8: 2674–2683, 2009.

[11] Park, Y.J., L.C. Chang, J.F. Liang, C. Moon, C.P. Chung, and V.C. Yang. Nontoxic membrane translocation peptide from protamine, low molecular weight protamine (LMWP), for enhanced intracellular protein delivery: *in vitro* and *in vivo* study. *FASEB Journal*, 19: 1555–1557, 2005.

[12] Choi, Y.S., J.Y. Lee, J.S. Suh, Y.M. Kwon, S.J. Lee, J.K. Chung, D.S. Lee, V.C. Yang, C.P. Chung, and Y.J. Park. The systemic delivery of siRNAs by a cell penetrating peptide, low molecular weight protamine. *Biomaterials*, 31: 1429-1443, 2010.

[13] Hauser, P.V., J.W. Pippin, C. Kaiser, R.D. Krofft, P.T. Brinkkoetter, K.L. Hudkins, D. Kerjaschki, J. Reiser, C.E. Alpers, and S.J. Shankland. Novel siRNA delivery system to target podocytes *in vivo*. *PLoS One*, 5: e9463, 2010.

12

Dendrimers

12.1 Introduction

Dendrimers (from the Greek "Dendron" which means tree, and "meros" means part) are branched, multivalent molecules with a well defined structure (Figure 12.1). These molecules were initially reported by Vögtle *et al.* as "cascade molecules" and later Tomalia *et al.* defined these molecules as dendrimers. These macromolecules are globular and nanometer in size with a distinct molecular architecture distinguishable into three different domains: (i) A central core which is either a single atom or a group of atoms having at least two chemical functionalities that facilitate the linkage of the branches; (ii) Branches emerging from the core, comprising repeat units with at least one junction of branching, whose repetition is organized in a geometric progression that results in a series of radially concentric layers termed generations (G); (iii) Terminal functional groups, present in the periphery of the macromolecule, which determines their nucleic acid complexation or drug entrapment capability. The presence of large number of terminal groups favors interactions with solvents, surfaces or other molecules. Generally, dendrimers possess high solubility, reactivity and binding properties.

Dendrimers can be synthesized gradually by a stepwise method of synthesis with a well defined size and structure with a comparatively low polydispersity index (PDI). Furthermore, dendrimer chemistry is quite amenable allowing synthesis of a wide range of molecules with different functionalities. Synthetic strategies for dendrimers are based on two conceptually different strategies, the divergent and the convergent approaches (Figure 12.2). In the

S. Nimesh and R. Chandra,
Theory, Techniques and Applications of Nanotechnology in Gene Silencing, 139–153.

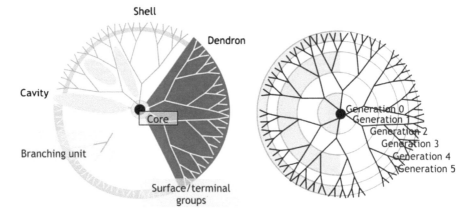

Fig. 12.1 Structure of Dendrimers.

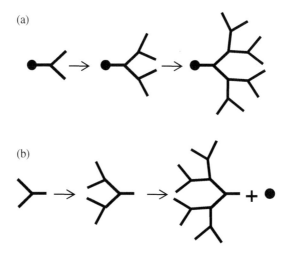

Fig. 12.2 Dendrimer synthetic approaches. (a) divergent synthesis, (b) convergent synthesis.

divergent method, the dendrimer is synthesised starting from the multifunctional core to buildup one monomer layer, or generation, at the time. The core molecule reacts with monomer molecules containing one reactive group and two (or more) inactive groups. In the convergent approach, the dendrimer is also built up layer after layer, but this time starting from the end groups and terminating at the core. Here two (or more) peripheral branch subunits are reacted with a single joining unit which has two (or more) corresponding active sites and a distal inactive site.

The first study on the use of Starbust® PAMAM dendrimers using luciferase and β-galactosidase as reporter plasmids showed that dendrimers mediate high transfection efficiency in a variety of suspension and adherent cultured mammalian cells with G6 (NH_3 core) dendrimer having the best efficiency. Maximal transfection of luciferase gene was obtained using a dendrimer with diameter of 6.8 nm and a N/P ratio of 6 (amine of dendrimer to phosphate of DNA). When GALA, a water soluble, membrane-destabilizing peptide, was covalently attached to the dendrimer via a disulfide linkage, transfection efficiency of the 1:1 complex increased by 2–3 orders of magnitude. It was argued that PAMAM dendrimers have the ability to escape from the endosome owing to the internal dendrimer amine groups to buffer pH changes in the endosome by a mechanism similar to that described for PEIs.

12.2 Applications of Dendrimers in siRNA Delivery

The unique characteristics of dendrimers such as high chemical and structural homogeneity, high ligand, and functionality density allow them to carry therapeutics by interior encapsulation, surface adsorption, or chemical conjugation. Polycationic dendrimers based on PAMAM and poly(propylenimine) (PPI) have been extensively investigated as efficient gene and drug delivery vectors. Recently, reports have appeared describing their potential in siRNA delivery.

12.2.1 Poly(Amidoamine) Dendrimers

Polycationic PAMAM dendrimers possess primary amine groups on their surface, and tertiary amine groups inside. The primary amine groups favor nucleic acid binding, their compaction into nanoscale particles and subsequent cellular uptake, while the tertiary amino groups trigger proton sponge in endosomes and enhance the release of nucleic acids into the cytoplasm. A comparative study done for cellular delivery of antisense and siRNA oligonucleotides using PAMAM G5 dendrimer conjugated to Tat peptide, a CPP. PAMAM dendrimers or their conjugates were found to be more efficient in delivering antisense oligonucleotides than delivering siRNA. The effects of antisense or siRNA delivered either by unmodified PAMAM G5 dendrimer or Tat conjugated dendrimer on cell surface P-glycoprotein expression in viable NIH

3T3 MDR cells were evaluated using immunofluorescence and flow cytometry, and compared to the inhibitory effects of antisense and siRNA delivered by Lipofectamine 2000. The dendrimers were comparably effective in reducing P-glycoprotein expression while delivering a phosphorothioate antisense oligonucleotide as compared to Lipofectamine 2000. On the contrary, when siRNA was delivered, the cationic lipid was much more efficient than the dendrimers in promoting reduction of P-glycoprotein expression. Furthermore, conjugation with Tat peptide failed to enhance delivery efficiency.

Pioneering work was reported by Zhou *et al.* in developing flexible polycationic PAMAM dendrimers for siRNA delivery. Triethanolamine was used as the core of the dendrimers and the branching units started 10 successive bonds away from the central nitrogen resulting in less densely packed branching units and end groups than that of the commercially available PAMAM dendrimers. Higher-generation dendrimers having structural features efficiently delivered siRNA and thus induced gene silencing in cell culture. The complexation ability of the dendrimer with siRNA increased with the increase in the generation number, due to the stronger interactions occurring between dendrimer and siRNA. The formation of stronger dendrimer-siRNA complexes at neutral pH and the efficient internalization of the siRNA molecules into the cytoplasm at higher N/P ratio led to greater gene silencing efficiency, presumably due to the dendrimer G7- mediated buffering of the endosomal cavity. The best gene silencing ($\sim$80%) was observed with G7 at N/P ratio of 10:20 and siRNA concentration of 100 nM. Toxicity studies revealed that the delivery step was relatively safe for the cells; although some toxicity was observed for the highest siRNA concentration. Further, these flexible triethanolamine core PAMAM dendrimers were used to deliver an Hsp27 siRNA effectively into prostate cancer (PC-3) cells by forming stable nanoparticles with siRNA. The Hsp27 siRNA resulted in potent and specific gene silencing of heat-shock protein 27, an attractive therapeutic target in castrate-resistant prostate cancer. Silencing of the Hsp27 gene led to induction of caspase-3/7-dependent apoptosis and inhibition of PC-3 cell growth *in vitro*.

The potential of dendrimers for siRNA delivery was further investigated by use of PAMAM dendrimer (G3) conjugated with α-cyclodextrin (α-CDE) to deliver siRNA-targeting luciferase gene and compared with various commercially available transfection reagents. The ternary complexes of pGL3/siGL3/α-CDE showed potent RNAi effects with negligible cytotoxicity in comparison with the transfection reagents in various cells. The α-CDE strongly interacted with both pDNA and siRNA, and protected siRNA degradation by serum, compared to those of the transfection reagents. Fluorescent labeled siRNA delivered with α-CDE was uniformly distributed in the cytoplasm, whereas that of transfection reagents resided in both nucleus and cytoplasm in NIH3T3 cells. Furthermore, the binary complex of siRNA/α-CDE provided the significant RNAi effect in NIH3T3 cells transiently and stably expressing luciferase gene. To acquire more insights into the preparation of stable uniform nanoscale dendrimer-RNA complexes, Shen *et al.* studied various dendrimer-RNA complexes using AFM. The study demonstrated that effective construction of stable nanoscale and uniform dendrimer-RNA complexes depends critically on the size of the RNA molecule, the dendrimer generation and the N/P ratio. Larger RNA molecules, higher generations of dendrimers and higher N/P ratios lead to the formation of stable, uniform nanoscale dendrimer-RNA complexes.

To circumvent the cytotoxcity of PAMAM-NH$_2$ dendrimers, Patil *et al.* derivatized the surface functionality and synthesized PAMAM-NHAc dendrimers. In this, dendrimer surface amine groups were modified with acetyl group and internal tertiary nitrogens were quaternized. The physicochemical properties and siRNA delivery efficacy of this dendrimer were compared with PAMAM-NH$_2$ dendrimer and with an internally quaternized charged, hydroxyl-terminated QPAMAM-OH dendrimer. AFM studies revealed that QPAMAM-OH and QPAMAM-NHAc, formed well-condensed, spherical particles with siRNA, while PAMAM-NH2 resulted in the formation of nanofibers. The modification of surface amine groups to amide significantly reduced cytotoxicity of dendrimers with QPAMAM-NHAc dendrimer showing the least toxicity. Confocal microscopy studies done on A2780 human ovarian cancer cells demonstrated that PAMAM-NH$_2$-siRNA and QPAMAM-OH-siRNA complexes did not penetrate into cellular cytoplasm within 80 min. of incubation time. On the contrary, the QPAMAM-NHAc-siRNA complexes were internalized by the cells and distributed uniformly inside the cytoplasm and nuclei by the end of the incubation period. Therefore, it was proposed that apart from the reduced non-specific interactions between dendrimer and cell membrane, the amide groups on the surface of the dendrimer might also play a crucial role by facilitating hydrogen bonding with cell membranes to enhance the cellular uptake of the complex.

To further address the issues of cytotoxicity, the primary amines of PAMAM-NH$_2$ were acetylated to different extents using acetic anhydride. Dendrimers with up to 60% of primary amines acetylated resulted in the formation of ~200 nm complexes with siRNA. With the increase in amine acetylation, polymer cytotoxicity reduced in U87 cells, alongwith enhanced dissociation of dendrimer-siRNA complexes. Acetylation of dendrimers reduced the cellular delivery of siRNA which correlated with a reduction in the buffering capacity of dendrimers upon amine acetylation. Confocal microscopy suggested that escape from endosomes is a major barrier to siRNA delivery in this system. It was concluded that a modest fraction (~20%) of primary amines of PAMAM can be modified while maintaining the siRNA delivery efficiency of unmodified PAMAM, but higher degrees of amine neutralization reduced the gene silencing efficiency of PAMAM-siRNA delivery vectors.

In order to improve the siRNA delivery efficiency of QPAMAM-OH dendrimer, a synthetic analog of Luteinizing hormone-releasing hormone (LHRH) as cancer-targeting moiety was tagged onto the dendrimer. The impact of two independent factors (degree of quaternization and LHRH-targeting ligand) was also examined. The complexes formed by using QPAMAM-OH and QPAMAM-OH-LHRH dendrimers exhibited a decrease in the particle size as the charge ratio was increased. Also, AFM at N/P ratio 3 revealed formation of well-condensed spherical particles. The cyto-toxicity evaluation of each dendrimer at various concentrations by the MTT assay revealed that the neutral surface and internally charged dendrimers QPAMAM-OH and QPAMAM-OH-LHRH possess cell viability above 90% even at relatively high concentrations up to 12.5 μM. Uptake studies showed that both the dendrimer and dendrimer-siRNA complex were internalized by A2780 cells and distributed uniformly in cellular cytoplasm and nucleus. The comparison of cellular internalization of siRNA delivered by dendrimers with different degree of quaternization showed that dendrimers with lower degree of quaternization (20–30%) delivered siRNA more efficiently when compared with those with higher degree (70–85%) of quaternization. The gene knockdown efficiency of siRNA-targeting BCL2 gene, delivered by non-targeted QPAMAM-OH and targeted QPAMAM-OH-LHRH dendrimer demonstrated significant suppression of the expression of BCL2 gene in A2780 cells employing targeted dendrimer. The study proposes that the degree of quaternization to some extent is important for the cellular uptake but not sufficient to achieve a gene silencing effect.

Perez *et al.* investigated the use of "not flexible" different size ethylendiamine (EDA) core PAMAM dendrimers to form complexes with siRNA, with emphasis on the influence of ionic strength of the preparation media on size, relative binding affinity and zeta potential of complexes, and its relation to their cellular uptake and silencing activity. It was observed that all the dendrimers formed complexes with siRNA, with dependence on the size of the complexes on the ionic strength of the media. The size of the complexes was smaller in the absence of NaCl than in presence of NaCl (30–130 nm, +25 mV zeta potential versus several μm-800 nm, 0 zeta potential, respectively). Also, both the uptake and inhibition of EGFP expression in the cell culture was dependent on the complex size. Complexes prepared in NaCl containing medium were poorly internalized and produced a basal activity on phagocytic (J-774-EGFP) cells, being inactive on non-phagocytic cells (T98G-EGFP). However, the smaller complexes prepared in NaCl lacking medium were efficiently internalized, exhibiting the highest inhibition of EGFP expression at 50 nM siRNA (non-cytotoxic concentration). Notably, siRNA-G7 complexes produced the highest inhibition of EGFP expression both in T98G-EGFP (35%) and J-774-EGFP (45%) cells, in spite of providing a lower protection to siRNA against RNase A degradation.

Further, to characterize the self-assembly process between siRNA and G7 PAMAM dendrimers resulting in dendriplexes, structural and calorimetric methods combined with molecular dynamics simulations were employed. Complexes with a size of 150 nm showed a decreasing size with increasing N/P ratio by DLS. At the molecular level, individual dendrimers studied by small-angle X-ray scattering (SAXS) showed no change in size upon siRNA binding, suggesting rigid sphere behavior. ITC demonstrated exothermic binding with a concentration dependent collapse of complexes. Both the experimentally determined ΔH of binding and size were in close accordance with the molecular dynamics simulations. This study demonstrated the unique complementarity of SAXS, ITC, and modeling for the detailed description of the molecular interactions between dendrimers and siRNA during dendriplex formation.

Furthermore, the self-assembly process between siRNA and different generation PAMAM dendrimers was elucidated and the resulting structures were characterized. The G4 and G7 displayed equal efficiencies for dendriplex aggregate formation, whereas G1 lacked this ability. Nanoparticle tracking analysis and DLS showed reduced average size and increased polydispersity at higher dendrimer concentration. The nanoparticle tracking analysis indicated that electrostatic complexation results in an equilibrium between different sized complex aggregates, where the center of mass depends on the dendrimer-siRNA ratio. ITC data suggested a simple binding for G1, whereas a biphasic binding was evident for G4 and G7 with an initial exothermic binding and a secondary endothermic formation of larger dendriplex aggregates, followed by agglomeration. The initial binding became increasingly exothermic as the generation increased, and the values were closely predicted by molecular dynamics simulations, which also demonstrated a generation dependent differences in the entropy of binding. The flexible G1 displayed the highest entropic penalty followed by the rigid G7, making the intermediate G4 the most suitable for dendriplex formation, showing favorable charge density for siRNA binding.

The high affinity interaction between RGD peptides and cancer related integrins has led to the widespread use of RGD peptide sequences as ligands for integrin-targeted drug and gene delivery applications. To exploit this strategy, Waite *et al.* derivatized PAMAM G5 dendrimers by incorporation of cyclic RGD-targeting peptides and evaluated their ability to associate with siRNA and mediate siRNA delivery to U87 malignant glioma cells. PAMAM-RGD conjugates efficiently complexed with siRNA to form complexes of ~200 nm in size. Modest siRNA delivery was observed in U87 cells using either PAMAM or PAMAM-RGD conjugates. PAMAM-RGD conjugates prevented the adhesion of U87 cells to fibrinogen-coated plates, in a manner that depends on the number of RGD ligands per dendrimer. The delivery of siRNA to three-dimensional multicellular spheroids of U87 cells was enhanced using PAMAM-RGD conjugates compared to the native PAMAM dendrimers, presumably by interfering with integrin-ECM contacts present in a three-dimensional tumor model.

Arginine-PAMAM esters (e-PAM-R) contain ester bonds between the surface hydroxyl group of PAMAM-OH and the carboxylate groups of arginine. e-PAM-R was found to be readily degradable under physiological conditions (pH 7.4, 37°C): More than 50% of the grafted amino acids were hydrolyzed within 5 h. Further, this vector was evaluated for siRNA delivery in primary cortical cultures in normal and ischemic rat brain. Localization of florescence-tagged siRNA revealed that siRNA was well delivered into the nucleus, cytoplasm and along the processes of the neuron. The e-PAM-R-siRNA complex-mediated target gene inhibition was observed in over 40% of cells and was persistent for over 48 h. The potential use of e-PAM-R was demonstrated by a gene knockdown after transfecting high mobility group box-1 (HMGB1, a novel cytokine-like molecule) siRNA into H_2O_2- or NMDA-treated primary cortical cultures. In these cells, HMGB1 siRNA delivery successfully reduced both basal and H_2O_2- or NMDA induced HMGB1 levels, and resulting in significant suppression of neuronal cell death. Furthermore, e-PAM-R successfully delivered HMGB1 siRNA into the rat brain, blocking HMGB1 expression in over 40% of neurons and astrocytes of the normal brain. Moreover, e-PAM-R-mediated HMGB1 siRNA delivery notably reduced infarct volume in the post-ischemic rat brain, generated by occluding the middle cerebral artery for 60 min. This study proposes e-PAM-R as a novel biodegradable non-viral gene carrier, capable of transfecting siRNA into primary neuronal cells and in the brain to mediate gene silencing.

To address major issues in siRNA delivery, such as poor cellular uptake, low endosomal escape, and facile enzymatic degradation, a novel triblock nanocarrier, PAMAM-PEG-PLL, was designed to combine individual features of PAMAM dendrimer, PEG, and PLL. PAMAM dendrimer provides tertiary amines for endosomal escape; PEG covers up siRNA, protecting it from enzymatic degradation, while PLL offers cationic amine groups for electrostatic interaction with negatively charged siRNA. The results suggest that the polyplexes resulted from the conjugation of siRNA, and the proposed nanocarriers were effectively taken up by cancer cells and induced the knockdown of the target BCL2 gene. In addition, triblock nanocarrier-siRNA polyplexes showed excellent stability in human plasma.

Incorporation of neutral groups into the periphery of PAMAM dendrimers can reduce cytotoxicity, although the cellular uptake of the dendrimer-siRNA complex may also be reduced. Additionally, increasing flexibility and/or generation number appears to improve gene knockdown effects. With the aim to evaluate the capability of delivering siRNA, triazine dendrimers, which have been successfully used for DNA transfection were studied. The smallest complexes of 105 and 103 nm, respectively, were formed with G2-3 and G2-5, which are comparable in size to PEI complexes of 103 nm. G2-1 dendriplexes exhibited a zeta potential of 8.5 mV, which is comparable to PEI complexes of 8.1 mV, indicating the presence of protonated amines on the surface of these complexes. The structure of the backbone was found to significantly influence siRNA transfection efficiency, with rigid, second generation dendrimers displaying higher gene knockdown than the flexible analogues while maintaining less off-target effects than Lipofectamine. Also, the rigid second generation dendrimers, with either arginine like exteriors or peripheries containing hydrophobic functionalities mediated the most effective gene knockdown, thus showing that dendrimer surface groups also affect transfection efficiency. Moreover, these two most effective dendriplexes were stable in circulation upon intravenous administration and showed passive targeting to the lung. Both dendriplex formulations were taken up into the alveolar epithelium, making them promising candidates for RNAi in the lung.

Dendrosomes are novel vesicular, spherical, supramolecular entities wherein the dendrimer-nucleic acid complex is entrapped within a lipophilic shell. They possess negligible hemolytic toxicity and higher transfection efficiency, and are better tolerated *in vivo* than the dendrimers. To explore the potential of dendrosomes, Dutta *et al.* delivered siRNA-targeting E6-E7 proteins of cervical cancer cells, *in vitro*. The formulation 4D100 (dendrimer-siRNA complex) displayed the highest GFP knockdown but was also found to be highly toxic to cells. In the final formulation, 4D100 was encapsulated into dendrosomes in order to mask these toxic effects. The optimized dendrosomal formulation (DF), DF3 was found to possess a siGFP-entrapment efficiency of 49.76% ± 1.62%, vesicle size of 154 ± 1.73 nm, and zeta potential of +3.21 ± 0.07 mV. The GFP knockdown efficiency of DF3 was found to be almost identical to that of 4D100, but the former was completely non-toxic to the cells. DF3 containing siRNA against E6 and E7 was found to knock down the target genes considerably, as compared with the other formulations tested. These results suggest of dendrosomes being potential vector for siRNA delivery and that a suitable targeting strategy could be useful for *in vivo* applications.

12.2.2   Poly(Propylenimine) Dendrimers

Few reports have been published which propose PPI dendrimers as siRNA delivery agents. In order to overcome low internalization of siRNA with its limited stability in blood, the PPI dendrimer was formulated for siRNA delivery systems through layer-by-layer modification approach including caging of siRNA-PPI (generation 5) dendriplex with a dithiol containing crosslinker molecules followed by coating with PEG polymer. A synthetic analog of LHRH peptide was conjugated to the distal end of PEG polymer to achieve targeted delivery of siRNA nanoparticles specifically to the cancer cells. The integrated attributes were effective in imparting serum resistance, increased stability in biological liquids, tumor-specific targeting, effective penetration into cancer cells, and accumulation of delivered siRNA in the cytoplasm while preserving the gene silencing ability. Further, it was demonstrated that the higher generations of PPI dendrimers (G4 and G5) were more effective in forming discrete nanoparticles upon complexation with siRNA in comparison to lower generations of dendrimers (G2 and G3). Larger size and higher positive charge density of G5 dendrimer elicited higher toxicity whereas G4 dendrimer displayed maximum efficacy with respect to siRNA nanoparticles formation, intracellular siRNA internalization, and sequence specific gene silencing. The formulated PPI-siRNA dendrimer complexes exhibited dramatic enhancement in siRNA cellular internalization and marked knockdown of targeted mRNA expression in A549 human lung cancer cells.

In an interesting and novel approach Chen *et al.* efficiently packaged and deliver siRNA with low generation (G3) PPI dendrimers by using AuNPs as a "labile catalytic" packaging agent. The AuNPs facilitated low generation dendrimers to package siRNA into discrete nanoparticles but more importantly, the AuNPs could be selectively removed from the siRNA complex solution without influencing the integrity of the siRNA complexes. Moreover, G3 PPI modified with AuNPs could be internalized into cancer cells, and the delivered siRNAs could efficiently silence their target mRNA. The efficiency of mRNA silencing by this approach was even superior to higher generation dendrimers (G5 PPI).

12.2.3 Carbosilane Dendrimers

Carbosilane dendrimers comprises interior carbon–silicon bonds that are slowly hydrolyzed when the dendrimer dissolves in water. This results in a gradual release of the exterior branches and their payload. The bulk of this time-dependent release has been estimated to be between 4 and 24 h. So, in order to bind and protect the siRNA while in transit to the target cells, and facilitate its transfection into the cytoplasm, and release the siRNA once inside the cells, carbosilane dendrimers (CBS) were investigated. Initial stability studies showed that CBS bind siRNA via electrostatic interactions. Dendrimer-bound siRNA was found to be resistant to degradation by RNase. Cytotoxicity assays of CBS-siRNA dendriplexes with peripheral blood mononuclear cells (PBMC) and the lymphocytic cell line SupT1 revealed a maximum safe dendrimer concentration of 25 μg/ml. Flow cytometry and confocal microscopy data revealed that lymphocytes were efficiently transfected by fluorochrome-labeled siRNA either naked or complexed with CBS. Dendriplexes with N/P ratio of 2 were determined to have the highest transfection efficiency while maintaining a low level of toxicity in these systems including hard-to-transfect HIV-infected PBMC. Finally, CBS-siRNA dendriplexes were shown to silence GAPDH expression and reduce HIV replication in SupT1 and PBMC. These findings suggests of feasibility of utilizing dendrimers such as CBS to deliver and transfect siRNA into lymphocytes thus allowing the use of RNAi as a potential alternative therapy for HIV infection.

Recently, the formation of dendriplexes between amino-terminated carbosilane dendrimers NN8, NN16 and antiHIV siRNA (siGAG1) has been investigated. The complexes were studied by circular dichroism, fluorescence, and zeta potential, the size of nanoparticles formed was estimated by DLS. The circular dichroism study showed that siGAG1 was fully bound by dendrimers at charge ratio of 1:3. Both the dendrimers bound to siRNA displaced ethidium bromide from its sites of intercalation. The NN16 dendrimer interacted more strongly with siRNA than NN8. On the basis of zeta potential data, the charge ratio of NN16/siGAG1 was calculated as 3:1 and that of NN8/siGAG1 as 4:1. The size of complexes ranged between 270 and 300 nm at a charge ratio of between 4:1 and 8:1 for the NN8/siRNA complex, and between 300 and 370 nm for the NN16/siRNA complex.

12.3   Future Perspectives

The general requirement for the development of synthetic delivery systems is a relatively non-toxic cationic material which can form a complex with siRNA in size range of few nanometers which show retardation in agarose gels, prevent binding of ethidium bromide, and protect the siRNA from degrading nucleases. The next step generally consists of testing the ability of complexes formed at various N/P ratios to increase manipulate gene silencing efficacy in a number of cell lines in the presence or absence of serum components. When the system shows effective knockdown efficiency compared to currently used standard reagents they are considered for *in vivo* testing, using reporter gene expression in various organs as the readout. Dendrimer-based transfection agents and specifically the fractured PAMAM dendrimers (Superfect®) are among the best and most widely applicable transfection agents and has become a standard tool for many cell and molecular biologists. The initial investigation of dendrimers for siRNA delivery has shown promising results which have been well translated to *in vivo* models. However, there is wide scope to improve upon the transfection conditions thereby enhancing the efficacy of delivered siRNA.

12.4 Conclusions

The dendrimer-based delivery systems have shown considerable promise as tools for further development of genetic therapies. While most of the applications so far have focused on the use of dendrimer-based vectors for local or *ex vivo* administration, some of recent studies have demonstrated that specifically PAMAM dendrimers possess properties which appear to make them particularly suited to systemic *in vivo* administration. Undoubtedly the obstacle of safe and efficient delivery of genetic medicine largely remains significant and the suitability of any siRNA delivery systems will always have to be matched with the clinical relevance.

References

[1] Buhleier, E., W. Wehner, and F. VÖGtle. "Cascade"- and "Nonskid-Chain-like" syntheses of molecular cavity topologies. *Synthesis*, 1978: 155–158, 1978.

[2] Tomalia, D.A., H. Baker, J. Dewald, M. Hall, G. Kallos, S. Martin, J. Roeck, J. Ryder, and P. Smith. A new class of polymers: Starburst-dendritic macromolecules. *Polymer Journal*, 17: 117–132, 1985.

[3] Hawker, C.J., and J.M.J. Frechet. Preparation of polymers with controlled molecular architecture. A new convergent approach to dendritic macromolecules. *Journal of the American Chemical Society*, 112: 7638–7647, 1990.

[4] Haensler, J., and F.C. Szoka. Polyamidoamine cascade polymers mediate efficient transfection of cells in culture. *Bioconjugate Chemistry*, 4: 372–379, 1993.

[5] Kang, H., R. DeLong, M. Fisher, and R. Juliano. Tat-conjugated PAMAM dendrimers as delivery agents for antisense and siRNA oligonucleotides. *Pharmaceutical Research*, 22: 2099–2106, 2005.

[6] Zhou, J., J. Wu, N. Hafdi, J.P. Behr, P. Erbacher, and L. Peng. PAMAM dendrimers for efficient siRNA delivery and potent gene silencing. *Chemical communications (Cambridge)*, 22: 2362–2364, 2006.

[7] Liu, X.X., P. Rocchi, F.Q. Qu, S.Q. Zheng, Z.C. Liang, M. Gleave, J. Iovanna, and L. Peng. PAMAM dendrimers mediate siRNA delivery to target Hsp27 and produce potent antiproliferative effects on prostate cancer cells. *ChemMedChem*, 4: 1302–1310, 2009.

[8] Tsutsumi, T., F. Hirayama, K. Uekama, and H. Arima. Evaluation of polyamidoamine dendrimer/[alpha]-cyclodextrin conjugate (generation 3, G3) as a novel carrier for small interfering RNA (siRNA). *Journal of Controlled Release*, 119: 349–359, 2007.

[9] Shen, X.C., J. Zhou, X. Liu, J. Wu, F. Qu, Z.L. Zhang, D.W. Pang, G. Quelever, C.C. Zhang, and L. Peng. Importance of size-to-charge ratio in construction of stable and uniform nanoscale RNA/dendrimer complexes. *Organic & Biomolecular Chemistry*, 5: 3674–3681, 2007.

[10] Patil, M.L., M. Zhang, S. Betigeri, O. Taratula, H. He, and T. Minko. Surface-modified and internally cationic polyamidoamine dendrimers for efficient siRNA delivery. *Bioconjugate Chemistry*, 19: 1396–1403, 2008.

[11] Waite, C.L., S.M. Sparks, K.E. Uhrich, and C.M. Roth. Acetylation of PAMAM dendrimers for cellular delivery of siRNA. *BMC Biotechnology*, 9: 38, 2009.

[12] Patil, M.L., M. Zhang, O. Taratula, O.B. Garbuzenko, H. He, and T. Minko. Internally cationic polyamidoamine PAMAM-OH dendrimers for siRNA delivery: Effect of the degree of quaternization and cancer targeting. *Biomacromolecules*, 10: 258–266, 2009.

[13] Perez, A.P., E.L. Romero, and M.J. Morilla. Ethylendiamine core PAMAM dendrimers/siRNA complexes as in vitro silencing agents. *International Journal of Pharmaceutical*, 380: 189200, 2009.

[14] Jensen, L.B., K. Mortensen, G.M. Pavan, M.R. Kasimova, D.K. Jensen, V. Gadzhyeva, H.M. Nielsen, and C. Foged. Molecular characterization of the interaction between siRNA and PAMAM G7 dendrimers by SAXS, ITC, and molecular dynamics simulations. *Biomacromolecules*, 11: 3571–3577, 2010.

[15] Jensen, L.B., G.M. Pavan, M.R. Kasimova, S. Rutherford, A. Danani, H.M. Nielsen, and C. Foged. Elucidating the molecular mechanism of PAMAM-siRNA dendriplex self-assembly: Effect of dendrimer charge density. *International Journal of Pharmaceutical*, in press, corrected proof.

[16] Waite, C.L., and C.M. Roth. PAMAM-RGD conjugates enhance siRNA delivery through a multicellular spheroid model of malignant glioma. *Bioconjugate Chemistry*, 20: 1908–1916, 2009.

[17] Nam, H.Y., H.J. Hahn, K. Nam, W.H. Choi, Y. Jeong, D.E. Kim, and J.S. Park. Evaluation of generations 2, 3 and 4 arginine modified PAMAM dendrimers for gene delivery. *International Journal of Pharmaceutical*, 363: 199–205, 2008.

[18] I.-D. Kim, C.-M. Lim, J.-B. Kim, H.Y. Nam, K. Nam, S.-W. Kim, J.-S. Park, and J.-K. Lee. Neuroprotection by biodegradable PAMAM ester (e-PAM-R)-mediated HMGB1 siRNA delivery in primary cortical cultures and in the postischemic brain. *Journal of Controlled Release*, 142: 422–430, 2010.

[19] Patil, M.L., M. Zhang, and T. Minko. Multifunctional triblock nanocarrier (PAMAM-PEG-PLL) for the efficient intracellular siRNA delivery and gene silencing. *ACS Nano*, 5: 1877–1887, 2011.

[20] Merkel, O.M., M.A. Mintzer, J. Sitterberg, U. Bakowsky, E.E. Simanek, and T. Kissel. Triazine dendrimers as nonviral gene delivery systems: effects of molecular structure on biological activity. *Bioconjugate Chemistry*, 20: 1799–1806, 2009.

[21] Mintzer, M.A., O.M. Merkel, T. Kissel, and E.E. Simanek. Polycationic triazine-based dendrimers: Effect of peripheral groups on transfection efficiency. *New Journal of Chemistry*, 33: 1918–1925, 2009.

[22] Merkel, O.M., M.A. Mintzer, D. Librizzi, O. Samsonova, T. Dicke, B. Sproat, H. Garn, P.J. Barth, E.E. Simanek, and T. Kissel. Triazine dendrimers as nonviral vectors for in vitro and in vivo RNAi: The effects of peripheral groups and core structure on biological activity. *Molecular Pharmaceutics*, 7: 969–983, 2010.

[23] Dutta, T., M. Burgess, N.A.J. McMillan, and H.S. Parekh. Dendrosome-based delivery of siRNA against E6 and E7 oncogenes in cervical cancer. *Nanomedicine: Nanotechnology, Biology and Medicine*. 6: 463–470, 2010.

[24] Taratula, O., O.B. Garbuzenko, P. Kirkpatrick, I. Pandya, R. Savla, V.P. Pozharov, H. He, and T. Minko. Surface-engineered targeted PPI dendrimer for efficient intracellular and intratumoral siRNA delivery. *Journal of Controlled Release*, 140: 284–293, 2009.

[25] Chen, A.M., O. Taratula, D. Wei, H.-I. Yen, T. Thomas, T.J. Thomas, T. Minko, and H. He. Labile catalytic packaging of DNA/siRNA: control of gold nanoparticles "out" of DNA/siRNA complexes. *ACS Nano*, 4: 3679–3688, 2010.

[26] Weber, N., P. Ortega, M.I. Clemente, D. Shcharbin, M. Bryszewska, F.J. de la Mata, R. Gomez, and M.A. Munoz-Fernandez. Characterization of carbosilane dendrimers as effective carriers of siRNA to HIV-infected lymphocytes. *Journal of Controlled Release*, 132: 55–64, 2008.

[27] Shcharbin, D., E. Pedziwiatr, L. Chonco, J.F. Bermejo-Martín, P. Ortega, F.J. de la Mata, R. Eritja, R. Gómez, B. Klajnert, M. Bryszewska, and M.A. Muñoz-Fernandez. Analysis of interaction between dendriplexes and bovine serum albumin. *Biomacromolecules*, 8: 2059–2062, 2007.

[28] Shcharbin, D., E. Pedziwiatr, O. Nowacka, M. Kumar, M. Zaborski, P. Ortega, F. Javier de la Mata, R. Gómez, M.A. Muñoz-Fernandez, and M. Bryszewska. Carbosilane dendrimers NN8 and NN16 form a stable complex with siGAG1. *Colloids and Surfaces B: Biointerfaces*, 83: 388–391, 2011.

13

Cyclodextrin

13.1 Introduction

Cyclodextrins (CDs) are cyclic (α-1,4)-linked oligosaccharides of α-D-glucopyranose containing a central hydrophobic cavity and hydrophilic outer surface. Due to the absence of free rotation around the bonds that connect the glucopyranose units, the CDs are not perfectly cylindrical molecules but are toroidal or cone shaped. In this structural arrangement, the primary hydroxyl groups are located on the narrow side of the torus while the secondary hydroxyl groups are located on the wider edge (Figure 13.1). The most common CDs are α-CD, β-CD and γ-CD which consist of six, seven and eight glucopyranose units, respectively. Davie *et al.* reported the first example of linear cationic polymers containing β-CD in the polymer backbone for gene delivery. Generally, the CD containing cationic polymers showed lower cytotoxicity and efficient gene transfection in cell cultures. The most important characteristic of the cyclodextrin containing cationic polymer (CDP) gene delivery system is that the polyplexes formed between the polymers and DNA can be further modified by inclusion- complex-formation due to presence of large amounts of CD moieties . A number of cationic polymers such as linear and branched PEI, and PAMAM dentrimer, were modified by grafting CDs on the polymers, and studied for gene delivery. All the CDPs showed reduced cytotoxicity by grafting CD moieties. Uekama *et al.* systematically investigated CD-grafting onto PAMAM dendrimers with different generations (G2, G3, and G4), and attributed the enhanced gene transfection efficacy to increased cellular association and intracellular trafficking of plasmid DNA.

S. Nimesh and R. Chandra,
Theory, Techniques and Applications of Nanotechnology in Gene Silencing, 155–164.
© 2011 *River Publishers. All rights reserved.*

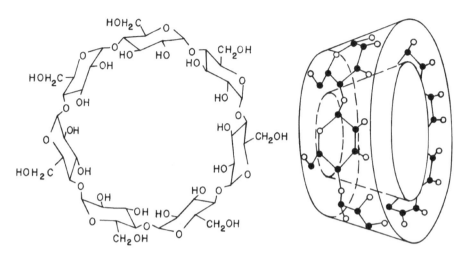

Fig. 13.1 (a) The chemical structure, (b) the toroidal shape of the β-cyclodextrin molecule.

Recently, a series of cationic star polymers were synthesized by conjugating multiple oligoethylenimine (OEI) arms onto α-CD core as non-viral gene delivery vectors. The molecular structures of the α-CD-OEI star polymers, which contained linear or branched OEI arms with different chain lengths ranging from 1–14 ethylenimine units, were characterized by using size exclusion chromatography and NMR techniques. The α-CD-OEI star polymers were studied in terms of their DNA binding capability, formation of nanoparticles with pDNA, cytotoxicity, and gene transfection in cultured cells. All the α-CD-OEI star polymers could inhibit the migration of pDNA on agarose gel via formation of complexes with pDNA, and the complexes formed nanoparticles with sizes ranging from 100–200 nm at N/P ratios of eight or higher. The star polymers displayed much lower *in vitro* cytotoxicity than that of branched PEI of MW 25 kDa. The α-CD-OEI star polymers showed excellent gene transfection efficiency in HEK 293 and COS-7 cells. Generally, the transfection efficiency increased with an increase in the OEI arm length. The star polymers with longer and branched OEI arms showed higher transfection efficiency. Although, the use of CDP is prevalent in pDNA delivery, it has also been successfully explored for siRNA delivery.

13.2   Cyclodextrins for siRNA Delivery

The multicomponent delivery system including short polycations containing cyclodextrins that provide low toxicity and enable assembly with the other components of the delivery system containing targeting ligands has been employed to deliver siRNA. The CDP self-assembles with siRNA to form colloidal particles ~50 nm in diameter and their terminal imidazole groups facilitates the intracellular trafficking and endosomal release of the nucleic acid. The CDP effectively protects the siRNA from nucleases degradation thus alleviating the need for chemical modification of the nucleic acid. The colloidal particles are stabilized for use in biological fluids by surface decoration with the PEG that occurs via inclusion complex formation between the terminal adamantane and the CDs; some of the PEG chains contain targeting ligands for specific interactions with cell-surface receptors. The complete formulation of the siRNA-containing particles was achieved by mixing the components together and allowing for the self-assembly.

In order to demonstrate a safe and systemic efficacy of non-virally delivered siRNA, Hu-Lieskovan *et al.* developed a mouse model of metastatic EFT in NOD/scid mice by tail-vein injections of EFT cells engineered to constitutively express luciferase. Further, they tested the ability of targeted, non-viral delivery of siRNA against EWS-FLI1 to safely limit bulk metastatic tumor growth and prevent establishment of bulk metastatic disease from microscopic metastatic disease. Complexes were prepared with siRNA and an imidazole-modified CDP; before addition to siRNA, CDP was mixed with an adamantane-PEG5000 (AD-PEG) conjugate at a 1:1 AD:β-CD (mol/mol) ratio. Targeted polyplexes also contained transferrin-modified AD-PEG (AD-PEG-transferrin) at a 1:1,000 AD-PEG-transferrin:AD-PEG (w/w) ratio. Imidazole-terminated CDP used to deliver siRNA sequence targeting the EWS-FLI1 breakpoint in TC71. Here, an EFT cell line that expresses the EWS-FLI1 fusion gene showed significant (>50%) reduction in EWS-FLI1 protein levels. The tumors of mice bearing luciferase expression, treated with the targeted, formulation siGL3-containing polyplexes showed a strong decrease (>90%) in luciferase signal two to three days after injection. The luciferase downregulation was transient. The luminescent signal increased daily thereafter. Further, three consecutive daily injections of the targeted, formulated siEFBP2 in mice with engraftment of TC71-LUC cells resulted in a decreased tumor signal with effect lasting two to three days. Moreover, assessment of the EWS-FLI1 expression in the tumors treated with two consecutive siEFBP2 formulations showed a 60% downregulation of EWS-FLI1 RNA level compared with control siRNA-treated tumors. Additionally, no abnormalities in interleukin-12 and IFN-A, liver and kidney function tests, complete blood counts, or pathology of major organs are observed from long-term, low-pressure, low-volume tail-vein administrations. This study provides strong evidence for the safety and efficacy of this targeted, non-viral siRNA delivery system.

In another study Bartlett *et al.* investigated the physicochemical and biological characterization of the CDP delivery system and its formulation with nucleic acids. These polycation-nucleic acid complexes can be manipulated by formulation conditions to yield particles with sizes ranging from 60 to 150 nm, zeta potentials from 10 to 30 mV, and MW from $\sim 7 \times 10^7$ to 1×10^9 g mol^{-1} as determined by light scattering techniques. Inclusion complexes formed between AD-containing molecules and the β-CD molecules enable the modular attachment of PEG (AD-PEG) conjugates for steric stabilization and targeting ligands (AD-PEG-transferrin) for cell-specific targeting. A 70 nm particle can contain $\sim$10,000 CDP polymer chains, $\sim$2,000 siRNA molecules, $\sim$4,000 AD-PEG5000 molecules, and $\sim$100 AD-PEG5000-Tf molecules; this represents a significant payload of siRNA and a large ratio of siRNA to targeting ligand (20:1). The particles efficiently protect the nucleic acid payload from nuclease degradation, do not aggregate at physiological salt concentrations, and cause minimal erythrocyte aggregation and complement fixation at the concentrations typically used for *in vivo* application. Uptake experiments show that the transferrin-targeted particles display enhanced affinity for the transferrin receptor through avidity effects (multi-ligand binding). Functional efficacy of the delivered pDNA and siRNA is demonstrated through luciferase reporter protein expression and knockdown, respectively. Co-delivery of a luciferase-expressing plasmid and either a control or luciferase-targeting siRNA was used to demonstrate the ability of the particles to deliver functional pDNA and siRNA. Cells that received CDP-Imidazole particles containing the plasmid and siRNA against luciferase had luciferase activity that was $\sim$50% lower than cells that received CDP-Imidazole particles with either the plasmid alone or the plasmid plus a control siRNA.

Nanoparticles containing multiple targeting ligands can have multivalent binding to cell surfaces and deliver a larger payload of siRNA than molecular conjugates. Heidel *et al.* reported multi-dosing study of siRNA in non-human primates with a targeted linear CDP for systemic delivery. When administered to cynomolgus monkeys at doses of 3 and 9 mg siRNA/kg, the nanoparticles were well tolerated. At 27 mg siRNA/kg, elevated levels of blood urea nitrogen and creatinine were observed that are indicative of kidney toxicity. Mild elevations in alanine amino transferase and aspartate transaminase at this dose level indicated that the liver was also affected to some extent. Analysis of complement factors does not reveal any changes that are clearly attributable to dosage with the nanoparticle formulation. Detection of increased IL-6 levels in all animals at 27 mg siRNA/kg and increased IFN-α in one animal indicate that this high dose level produces a mild immune response. Overall, no clinical signs of toxicity clearly attributable to treatment were observed. The multiple administrations spanning a period of 17–18 days enabled assessment of antibody formation against the human Tf component of the formulation. Low titers of anti-transferrin antibodies were detected, but this response was not associated with any manifestations of a hypersensitivity reaction upon read-ministration of the targeted nanoparticle. This study suggests that multiple, systemic doses of targeted nanoparticles containing unmodified siRNA can safely be administered to non-human primates.

Studies using nanoparticles to deliver chemotherapeutics or siRNA demonstrated that attachment of cell-specific targeting ligands to the surface of nanoparticles results in enhanced potency relative to non-targeted formulations. Bartlett *et al.* used positron emission tomography (PET) and bioluminescent imaging to quantify the *in vivo* biodistribution and function of nanoparticles formed with CDP and siRNA. Incorporation of 1,4,7,10-tetraazacyclododecane-1,4,7,10-tetraacetic acid to the 5' end of the siRNA molecules allows labeling with ^{64}Cu for PET imaging. Bioluminescent imaging of mice bearing luciferase-expressing Neuro2A subcutaneous tumors before and after PET imaging enables correlation of functional efficacy with biodistribution data. Although both non-targeted and transferrin-targeted siRNA nanoparticles exhibit similar biodistribution and tumor localization by PET, transferrin-targeted siRNA nanoparticles reduced tumor luciferase activity by ~50% relative to non-targeted siRNA nanoparticles 1 d after injection. Compartmental modeling showed that the primary advantage of targeted nanoparticles is associated with processes involved in cellular uptake in tumor cells rather than overall tumor localization. Hence, optimization of internalization may therefore be major factor for the development of effective nanoparticles-based targeted therapeutics.

Despite the extensive literature base demonstrating the therapeutic potential of siRNA, there has been much less attention devoted to issues surrounding siRNA-based treatments, including dosing schedule design. Bartlett *et al.* designed a study to address the issues of relevance for siRNA nanoparticle delivery by investigating the functional impact of tumor-specific targeting and dosing schedule. The investigations were performed using an experimental system involving a syngeneic mouse cancer model and a theoretical system based on mathematical model of siRNA delivery and function. The A/J mice bearing subcutaneous Neuro2A tumors $\sim$100 mm^3 in size were treated by intravenous injection with siRNA-containing nanoparticles formed with CDP. Three consecutive daily doses of transferrin-targeted nanoparticles carrying 2.5 mg/kg of two different siRNA sequences targeting the M2 subunit of ribonucleotide reductase (RRM2) slowed tumor growth, whereas non-targeted nanoparticles were significantly less effective when given at the same dose. Moreover, administration of the three doses on consecutive days or every three days did not lead to statistically significant differences in tumor growth delay. Mathematical model calculations of siRNA-mediated target protein knockdown and tumor growth inhibition were used to elucidate possible mechanisms to explain the observed effects and to provide guidelines for designing more effective siRNA-based treatment regimens regardless of delivery methodology and tumor type. It was deduced that once sufficient siRNA has been delivered to inhibit the growth of a given tumor cell, there is no advantage in delivering more siRNA to the same region of cells. This is particularly important for therapeutic siRNAs which act to arrest cell growth or elicit cell death, since a threshold may exist beyond which further knockdown no longer achieves any advantage (i.e., the cell is already growth-arrested or dying). In these situations, multiple doses may not be needed for any given cell. On the other hand, multiple doses might be important if new cells are reached that either have not internalized any siRNA or have not internalized sufficient siRNA to pass beyond the threshold required for the phenotypic effect such as cell death. The duration of target knockdown after siRNA treatment is an important factor to consider when designing treatments. These simulations demonstrate that the reduction in cell growth rate can lead to longer target knockdown because of the reduced dilution from cell division.

To identify transcriptionally-regulated pro-apoptotic genes in sepsis and to determine whether RNAi is targeting those genes individually or in combination could prevent sepsis-induced lymphocyte apoptosis targeted CDP was used for intracellular delivery of siRNAs. Using quantitative reverse transcription polymerase chain reaction (qRT-PCR) of isolated splenic CD4 T and B cells, it was determined that Bim and PUMA, two key cell death proteins, are markedly up-regulated during sepsis. Lymphocytes have reported to be difficult to transfect with siRNA. Therefore the authors employed a novel, CDP transferrin receptor (TFR)-targeted delivery vehicle to co-administer siRNA to Bim and PUMA to mice immediately after cecal ligation and puncture. Anti-apoptotic siRNA-based therapy markedly decreased lymphocyte apoptosis and prevented the loss of splenic CD4 T and B cells. Flow cytometry confirmed *in vivo* delivery of siRNA to CD4 T and B cells and also demonstrated decreases in intracellular Bim and PUMA protein. It was conclude that Bim and PUMA are two critical mediators of immune cell death in sepsis and novel CDP transferrin receptor-targeted siRNA delivery vehicle enables effective administration of anti-apoptotic siRNAs to lymphocytes and reverses the immune cell depletion.

In a recent published study by Davis *et al.*, phase I trial patients with melanomas were treated with siRNA therapy whose tumors had not responded to standard therapy. The siRNA used was directed against an established target for cancer therapy, the M2 subunit of ribonucleotide reductase (RRM2). The patients were administered doses of targeted nanoparticles on days 1, 3, 8, and 10 of a 21-day cycle by a 30 min intravenous infusion. The nanoparticles consist of a synthetic delivery system containing: (1) a linear CDP, (2) a human transferrin protein targeting ligand displayed on the exterior of the nanoparticle to engage TFR on the surface of the cancer cells, (3) a hydrophilic polymer PEG (used to promote nanoparticle stability in biological fluids), and (4) siRNA designed to reduce the expression of the RRM2. These nanoparticles (clinical version denoted as CALAA-01) have been shown to be well tolerated in multi-dosing studies in non-human primates. Tumor biopsies from melanoma patients obtained after treatment showed the presence of intracellularly localized nanoparticles in amounts that correlated with dose levels of the nanoparticles administered. Furthermore, a reduction was found in both the specific messenger RNA (RRM2 and the protein) levels when compared to predosing tissue. Most notably, the presence of an mRNA fragment was detected that demonstrates that siRNA-mediated mRNA cleavage occurs specifically at the site predicted for an RNAi mechanism from a patient who received the highest dose of the nanoparticles. It was suggested that that siRNA administered systemically to a human can produce a specific gene inhibition (reduction in mRNA and protein) by an RNAi mechanism of action.

Another very interesting strategy which does not require any endosomolytic agent but is endowed with excellent potential in directed delivery of siRNA, involves the utilization of PCI technology. PCI technology is pivoted around the preferential localization of photosensitizers at the endosomal and lysosomal membranes. Illumination of these PSs triggers the photochemical damage of the membrane followed by release of endosomal and lysosomal content. Importantly it is also capable of producing light-directed delivery of siRNA into the tissue of interest. Bøe *et al.* demonstrated the utilization of PCI-mediated light-directed delivery of CDP-siRNA complex. Optimization of carrier/cargo ratio and illumination dose provided 80% and 90% silencing in the siRNA samples treated with PCI compared with untreated control. On the contrary, only a 0%–10% silencing effect was detected in the siRNA samples without PCI treatment, demonstrating the potency of light-specific delivery of siRNA molecules. Further, time-lapse studies revealed maximum gene silencing at five hours after endosomal release substantiating rapid carrier decondensation for the CDP.

13.3 Future Perspectives

Owing to the requirement for cell-specific delivery of therapeutic siRNA, justifiable consideration needs to be given to the design and formulation of targeted, non-immunogenic and efficient delivery technologies. Cyclodextrin-based delivery strategies have shown enormous promise in the clinical trial studies done by Calando pharmaceuticals, and so far it is the only trial ongoing for treatment of solid tumors. Consequently, opportunities exist to develop further innovative delivery technologies to satisfy the enormous unmet need in this field. Hence, there is a call for design and development of better strategies for delivery of therapeutically relevant siRNA.

13.4 Conclusions

CALAA-01 is the first targeted delivery of siRNA in humans. It has been well compared with other non-targeted and targeted experimental and approved therapeutics. The treatment of patients with CALAA-01 ushers in a new era of targeted experimental therapeutics. Encouraging results obtained from this study paves the way towards development of other formulated siRNA therapeutics. Cyclodextrin based polycations have immense potential to be further manipulated for successful siRNA delivery.

References

[1] Gonzalez, H., S.J. Hwang, and M.E. Davis. New class of polymers for the delivery of macromolecular therapeutics. *Bioconjugate Chemistry*, 10: 1068–1074, 1999.

[2] Davis, M.E., and M.E. Brewster. Cyclodextrin-based pharmaceutics: Past, present and future. *Nature Reviews Drug Discovery*, 3: 1023–1035, 2004.

[3] Pack, D.W., A.S. Hoffman, S. Pun, and P.S. Stayton. Design and development of polymers for gene delivery. *Nature Reviews Drug Discovery*, 4: 581–593, 2005.

[4] Arima, H., F. Kihara, F. Hirayama, and K. Uekama. Enhancement of gene expression by polyamidoamine dendrimer conjugates with alpha-, beta-, and gamma-cyclodextrins. *Bioconjugate Chemistry*, 12: 476–484, 2001.

[5] Kihara, F., H. Arima, T. Tsutsumi, F. Hirayama, and K. Uekama. Effects of structure of polyamidoamine dendrimer on gene transfer efficiency of the dendrimer conjugate with alpha-cyclodextrin. *Bioconjugate Chemistry*, 13: 1211–1219, 2002.

[6] Wada, K., H. Arima, T. Tsutsumi, Y. Chihara, K. Hattori, F. Hirayama, and K. Uekama. Improvement of gene delivery mediated by mannosylated dendrimer/alpha-cyclodextrin conjugates. *Journal of Controlled Release*, 104: 397–413, 2005.

[7] Yang, C., H. Li, S.H. Goh, and J. Li. Cationic star polymers consisting of alpha-cyclodextrin core and oligoethylenimine arms as nonviral gene delivery vectors. *Biomaterials*, 28: 3245–3254, 2007.

[8] Davis, M.E., S.H. Pun, N.C. Bellocq, T.M. Reineke, S.R. Popielarski, S. Mishra, and J.D. Heidel. Self-assembling nucleic acid delivery vehicles via linear, water-soluble, cyclodextrin-containing polymers. *Current Medicinal Chemistry*, 11: 179–197, 2004.

[9] Hu-Lieskovan, S., J.D. Heidel, D.W. Bartlett, M.E. Davis, and T.J. Triche. Sequence-specific knockdown of EWS-FLI1 by targeted, nonviral delivery of small interfering RNA inhibits tumor growth in a murine model of metastatic ewing's sarcoma. *Cancer Research*, 65: 8984–8992, 2005.

[10] Bartlett, D.W., and M.E. Davis. Physicochemical and biological characterization of targeted, nucleic acid-containing nanoparticles. *Bioconjugate Chemistry*, 18: 456–468, 2007.

[11] Heidel, J.D., Z. Yu, J.Y. Liu, S.M. Rele, Y. Liang, R.K. Zeidan, D.J. Kornbrust, and M.E. Davis. Administration in nonhuman primates of escalating intravenous doses of targeted nanoparticles containing ribonucleotide reductase subunit M2 siRNA. *Proceedings of the National Academy of Sciences of the United States of America*, 104: 5715–5721, 2007.

[12] Bartlett, D.W., H. Su, I.J. Hildebrandt, W.A. Weber, and M.E. Davis. Impact of tumor-specific targeting on the biodistribution and efficacy of siRNA nanoparticles measured by multimodality *in vivo* imaging. *Proceedings of the National Academy of Sciences of the United States of America*, 104: 15549–15554, 2007.

[13] Bartlett, D.W., and M.E. Davis. Impact of tumor-specific targeting and dosing schedule on tumor growth inhibition after intravenous administration of siRNA-containing nanoparticles. *Biotechnology & Bioengineering*, 99: 975–985, 2008.

[14] P. Brahmamdam, E. Watanabe, J. Unsinger, K.C. Chang, W. Schierding, A.S. Hoekzema, T.T. Zhou, J.S. McDonough, H. Holemon, J.D. Heidel, C.M. Coopersmith, J.E. McDunn, and R.S. Hotchkiss. Targeted delivery of siRNA to cell death proteins in sepsis. *Shock*, 32: 131–139, 2009.

[15] J.D. Heidel, J.Y. Liu, Y. Yen, B. Zhou, B.S. Heale, J.J. Rossi, D.W. Bartlett, and M.E. Davis. Potent siRNA inhibitors of ribonucleotide reductase subunit RRM2 reduce cell proliferation *in vitro* and *in vivo*. *Clinical Cancer Research*, 13: 2207–2215, 2007.

[16] S.L. Boe, A.S. Longva, and E. Hovig. Cyclodextrin-containing polymer delivery system for light-directed siRNA gene silencing. *Oligonucleotides*, 20: 175–182, 2010.

14

Poly (D, L-Lactide-co-Glycolide)

14.1 Introduction

Although, cationic polymers (i.e. nucleic acids condensing agents) seem to be an excellent substitute for viral vectors, some limitations inherent in their cationic nature limits the application for systemic delivery. These include toxicity due to excess positive charge of the polymer, rapid clearance by the RES, inability of the complex to escape from the endosome/lysosome compartments in the cells, and lack of intracellular unpacking of the nucleic acid construct from the electrostatic complex. Non-condensing polymers forming nanosized vectors with either a neutral or net negative charge have been proposed for tissue and cell specific delivery that possess efficient transfection efficiency with improved cell viabilities. Plasmid DNA, siRNA or oligonucleotide is entrapped within the nanoparticles either by physical encapsulation of nucleic acid constructs within the matrix or via covalent coupling between polymer and nucleic acid bases (Figure 14.1). Encapsulation provides protection against nucleases and other plasma proteins during its movement across blood to the target site. Moreover, absence of positive charges on non-condensing vectors limits its recognition by the MPS and hence allows prolonged circulation.

Amongst various non-condensing vectors, PLGA has been suggested as a promising candidate for nucleic acids delivery. Studies proposed that PLGA nanoparticles are efficiently and rapidly internalized through a combination of specific and non-specific endocytic mechanisms. Upon internalization a significant amount of nanoparticles escape the lysosomal compartment and reach the cytosol. Surface charge reversal of PLGA nanoparticles from anionic to

S. Nimesh and R. Chandra,
Theory, Techniques and Applications of Nanotechnology in Gene Silencing, 165–176.

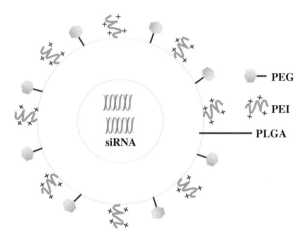

Fig. 14.1 Schematic representation of PLGA nanosphere. The surface is emodified by PEG coating and positive charge introduced by cationic polymer such as PEI.

cationic in the acidic pH of the lysosomes is regarded as an important aspect that enables PLGA nanoparticles to escape the lysosomal compartment. Following lysosomal escape, PLGA nanoparticles have been observed to retain in the cellular cytoplasm for more than two weeks, thereby slowly releasing their payload. Advantages such as efficient cellular uptake, rapid lysosomal escape and sustained intracellular drug release proposes PLGA nanoparticles as an attractive vector for gene silencing applications. Since PLGA degrades to body metabolites, i.e. lactic and glycolic acid, by hydrolysis of ester bonds, they have high biocompatibility, non-toxicity and ease of fabrication in different forms. Additionally, PLGA have been accepted by the US Food and Drug Administration for human use. PLGA is a copolymer of polylactic acid and polyglycolic acid synthesized in a wide range of MW by ring-opening polymerization of cyclic dimers, i.e. lactide and glycolide, in the presence of metal catalysts (Figure 14.2). Several modifications of PLGA have been reported for DNA or siRNA delivery, especially by copolymerization with other monomers. Depending on the ratio of lactide to glycolide used for the copolymerization, different forms of PLGA are available; these are usually identified on the basis of monomer's ratio used (e.g. PLGA 75:25 is a copolymer whose composition is 75% lactic acid and 25% glycolic acid). All PLGAs are amorphous and show a glass transition temperature in the range of 40–60°C.

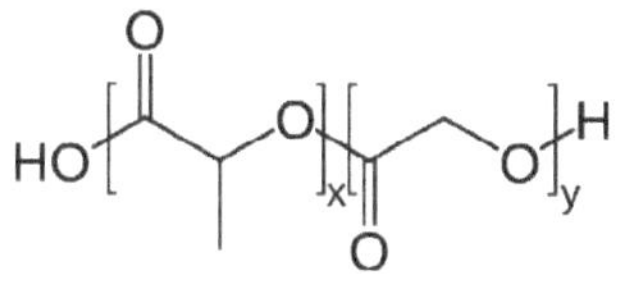

Fig. 14.2 Chemical structure of PLGA, X = number of lactic acid units and Y = number of glycolic acid units.

14.2 PLGA for siRNA Delivery

The versatility of PLGA allows chemical modification; several derivatives of PLGA with tailored properties have been designed to fit the delivery purposes. Sustained release of the encapsulated nucleic acids could be achieved by manipulating the degradation rate of PLGA. In one of initial studies, attempts were made to produce cationic PLGA particles using chitosan as a cationic surface modifier for the adsorption of siRNA. PLGA-chitosan particles obtained were positively charged with particle size between 400–1,000 nm depending on type, MW and concentration of chitosan as well as type of PLGA. Chitosan incorporated particles were much larger and had higher PDI depending on the MW of chitosan used. Particle size of PLGA-chitosan particles decreased with the decrease in chitosan MW. A better siRNA loading capacity was observed when a higher degree of "uncapped end groups" were used. The addition of trehalose has also been shown to stabilize these particles from severe aggregation induced by freeze–drying. It was found that physical properties of PLGA-chitosan particles and their siRNA binding capacity were highly influenced by preparation parameters. Also, the MTT assay showed that all the formulations of PLGA-chitosan particles were non-toxic to the cells giving percentage cell viability more than 95% for the tested particles to siRNA weight ratio (100:1 to 500:1).

On similar lines, to improve bioavailability, PLGA nanospheres surface was modified with chitosan and evaluated for *in vitro* siRNA delivery. The nanospheres produced by emulsion solvent diffusion method, loaded with siRNA were ∼300 nm and showed a positive zeta potential, while native-PLGA nanospheres were negatively charged. The surface morphology of siRNA-loaded PLGA nanospheres observed by SEM was spherical, smooth, and homogenous in distribution. The siRNA uptake studies with confocal microscopy indicated that siRNA-loaded chitosan modified PLGA nanospheres were more efficiently taken up by the cells than native ones. Luciferase gene silencing efficiencies of chitosan modified PLGA nanospheres assessed in A549 cells was higher and more prolonged than those of native-PLGA nanospheres and naked siRNA. Gene knockdown results were in perfect correlation with the uptake studies, likely to be caused by higher cellular uptake of chitosan modified PLGA nanospheres due to electrostatic interactions. In another study, uniform and small size chitosan modified PLGA nanoparticles were prepared by using 1.0% of PVA as emulsifier. The size of PLGA nanoparticles was 182 nm, which increased from 204 to 543 nm with the increase in chitosan concentration in the chitosan-PLGA nanoparticles. The zeta potential of the native PLGA nanoparticles was negative, -10.6 mV, whereas all chitosan-PLGA nanoparticles were positive, and zeta potential increased from $+16.9$ to $+31.2$ mV with the increase of chitosan coating concentration. *In vitro* gene silencing of PLGA nanoparticle was assessed by inhibiting GFP expression in HEK 293 cells. Maximal silencing effect ($63.3 \pm 5.6\%$), comparable with Lipofectamine was achieved. In a recent study, Zeng *et al.* also prepared chitosan incorporated into PLGA matrix nanoparticles (CS-PLGA NS) to improve pDNA loading efficiency and cellular uptake ability. CS-PLGA NS prepared by a spontaneous emulsion diffusion method was spherical with particle size of $\sim$60 nm. Further, CS-PLGA NS showed a positive zeta potential, while native-PLGA nanoparticles were negatively charged. These nanoparticles were evaluated for silencing the expression of HBV X and HBV S region, intended for treating HBV infectious diseases. In the HepG 2.2.15 cells transfected with CS-PLGA NS for 72 h, the levels of the HBV-X mRNA and HBV-S mRNA were noticeably reduced to 0.39 ± 0.02-fold and 0.49 ± 0.01-fold compared with control. The HBV-X gene silencing efficiency of CS-PLGA NS was higher than native PLGA nanoparticles and control, but showed a lower gene knockdown efficiency compared with Lipofectamine 2000.

In order to improve upon the properties of PLGA, branched biodegradable polyesters were designed by attaching hydrophilic, positively charged amine groups onto a hydrophilic water-soluble backbone consisting of PVA which was subsequently grafted with multiple PLGA side chains. The siRNA entrapped nanoparticles were prepared using a solvent displacement method that offers the advantage of forming small nanoparticles without using shear forces. The amine-modified-PVA-PLGA/siRNA nanoparticles formed were 150–200 nm in size with zeta potentials of $+15$mV to $+20$mV in PBS.

Degradation of the nanoparticles was seen within 4 h in PBS with sustained release of siRNA. *In vitro* knockdown of luciferase reporter gene was used to assess the potential of nanoparticles as siRNA carriers in a human lung epithelial cell line, H1299 luc. These nanoparticles achieved 80–90% knockdown of the luciferase gene with only 5 pmol antiluc siRNA, even after nebulization.

The conjugation of PLGA with poloxamers is a promising approach for the delivery of plasmid DNA. Luo *et al.* incorporated siRNA sequence of methyl-CpG binding domain protein 1 (MBD1) into PLGA:Poloxamer nanoparticles and tested the *in vitro* therapeutic effects on pancreatic cancer BxPC-3 cells. The nanoparticles prepared with or without encapsulation of plasmid containing the siRNA were found to be in range of 198 to 205 nm in size. The nanoparticles exhibited a biphasic pDNA release pattern in Tris-EDTA buffer (pH 7.4), characterized by a first initial rapid release (>30% of pDNA within the first day) followed by a slower, continuous release (90% released in 11 days). Gene knockdown studies revealed that MBD1 protein expression decreased from day 2 and completely disappeared at day 5. Apoptotic cells were observed by immunostaining for the presence of DNA fragments and the apoptosis rate was 24.19% in the nanoparticles transfected cells, which was higher than that of the control cells (4.79%).

To improve siRNA encapsulation in PLGA nanoparticles, cationic polymer, PEI was incorporated in the PLGA matrix. PLGA-PEI (40 kDa) nanoparticles were formulated using double emulsion-solvent evaporation technique. Particle size of nanoparticles was found to be dependent on PEI concentration. In the absence of PEI, PLGA nanoparticles had an effective hydrodynamic diameter of 280 nm which increased to 330 and 586 nm at PEI concentration of 0.5 and 500μg/30 mg PLGA respectively. Presence of PEI in PLGA nanoparticle matrix increased siRNA encapsulation by about two-fold and also improved the siRNA release profile. The efficacy of siRNA-loaded PLGA-PEI nanoparticles in silencing luciferase gene was investigated in both a stably transfected cell line (EMT-6 G/L) as well as in an inducible cell line (MDA-Kb2). The inhibition of luciferase expression was found to be greater and more sustained in cells treated with siRNA nanoparticles than in those transfected using a commercial transfecting agent Dharma FECT. In dexamethasone stimulated MDA-Kb2 cells, PLGA-PEI nanoparticles reduced the luciferase expression by about 70%, and this effect was sustained over three days. Further, quantitative studies indicated that presence of PEI in PLGA nanoparticles resulted in two-fold higher cellular uptake of nanoparticles while fluorescence microscopy studies showed that PLGA-PEI nanoparticles delivered the encapsulated siRNA in the cellular cytoplasm. It was postulated that both higher uptake and greater cytosolic delivery could have contributed to the gene silencing effectiveness of PLGA-PEI nanoparticles. Serum stability and lack of cytotoxicity further add to the potential of PLGA-PEI nanoparticles in gene silencing-based therapeutic applications.

In another study, PEI was conjugated to PLGA particles prepared by spontaneous modified emulsification diffusion method. Incorporation of PEI into PLGA particles with the PLGA to PEI weight ratio 29:1 was found to produce spherical and positively charged nanoparticles. Particle size of around 100 nm was obtained when 5% (m/v) PVA was used as a stabilizer. The PLGA-PEI nanoparticles completely shielded siRNA at N/P ratio 20:1 to provide protection against nuclease degradation. *In vitro* gene silencing activity of siRNA adsorbed onto PLGA-PEI nanoparticles were assessed in two different types of cells, namely HEK 293 and CHO K1 cell lines. In HEK 293 cells, the particles prepared using PVA with MW of 13-23 kDa, 99% of luciferase gene downregulation was observed at N/P ratio of 25:1 and the effect was slightly reduced to 96% and 87% for N/P ratio of 35:1 and 50:1, respectively. In contrast, in CHO K1 cell line, less pronounced gene silencing; only 54 and 15% was achieved for N/P ratio of 35:1 and 50:1. The MTT cell viability assays revealed of 15–25% of cell death in both cells lines. Working towards this direction, the surface of PLGA nanoparticles produced by an emulsion-diffusion method using benzyl alcohol modified with PEI utilising a cetyl derivative. The X-ray photoelectron spectroscopy (XPS) studies demonstrated 2.6 times higher surface presentation of amines using the cetyl derivative compared to noncetylated-PEI formulations (6.5 and 2.5% surface nitrogen, respectively). The modified particles were shown by spectroscopy, fluorescent microscopy and flow cytometry to bind and mediate siRNA delivery into the human osteosarcoma cell line U2OS and the murine macrophage cell line J774.1. Specific reduction in the anti-apoptotic oncogenes BCL-w in U2OS cells was achieved with particles containing cetylated-PEI (53%) with no cellular toxicity. Additionally, particles containing cetylated-PEI achieved 64% silencing of TNFα in J774.1 cells.

In a comparative study, PLGA nanoparticles were synthesized by nano-precipitation and solvent evaporation method; surface coated with cationic PEI (25 kDa) and chitosan. PEI decoration increased the particle size from 147.5 nm (for native PLGA) to 200.3 nm; while chitosan coating increased the particle to 260.5 nm. Nanoparticles coated with PEI fabricated with nano-precipitation method possessed smaller particle size than their counterpart probably because of the lower viscosity of PEI formulated water phase than the chitosan water phase. As expected, the zeta potential of the nanoparticles significantly changed from negative to positive value after being modified by PEI and chitosan. Complexation studies suggested that PLGA-PEI nanopar-ticles form complex with dsRNA when their ratio was above 6:1 (w/w) and that of PLGA-chitosan at 10:1 (w/w). PLC/PRF/5 human liver cell which pro-duce and secret surface antigen of the hepatitis B Virus (HBsAg) was used to assess the inhibition effect of S_2RNA delivered by PLGA nanoparticles. The expression levels of cells treated by nanoparticles/siRNA complex of PLGA/chitosan, PLGA/PEI were 42.50% and 24.26%, respectively. More-over, the cytotoxicity study of nanoparticles estimated by MTT assay revealed of cell viabilities above 90%.

A study was designed to prepare uniformly sized nanoparticles of PLGA with a high load of siRNA without including polycations. The siRNA was entrapped in the core of nanoparticles by the double emulsion solvent evapo-ration method. Physicochemical studies revealed of siRNA-loaded nanoparti-cles of < 300 nm, PDI < 0.2, zeta potential -40mV. Encapsulation efficiency of up to 57% was achieved by adjusting the inner water phase volume, the PLGA concentration, the first emulsification sonication time, and stabiliza-tion of the water-oil interface with serum albumin. The integrity of siRNA in the nanoparticles was evaluated by extracting siRNA from dissolved PLGA nanoparticles, followed by gel electrophoresis. siRNA appeared intact during the preparation methodology and critical steps such as sonication and freeze–drying had no influence on the integrity of the siRNA. Further, optimization of experimental conditions involving PLGA concentration and the volume ratio yielded encapsulation efficiency as high as 70%. The improved encapsulation was attributed to increased viscosity of the oil phase at high PLGA concentra-tion, which stabilizes the primary emulsion and reduces siRNA leakage to the outer water phase. Furthermore, the PLGA matrix protected siRNA against nuclease degradation provided a burst release of surface-localized siRNA followed by a triphasic sustained release for two months.

Hybrid systems formed by the incorporation of polycations into polymeric nanospheres are strategies to promote nucleic acid binding capacity and endosomal escape of solid nanoparticles whilst improving stability and lowering the toxicity associated with polycations. The disruption of tumor-induced hyperactive phosphorylated-STAT3, a transcription factor, in dendritic cells is considered as an attractive strategy for cancer immunotherapy. Alshamsan *et al.* investigated the influence of encapsulation of STAT3 siRNA/PEI and PEI-Stearic acid (PEI-StA) polyplexes in PLGA nanoparticles for STAT3 knockdown. Nanoparticles of PLGA containing siRNA polyplexes of PEI (PLGA-P) and PEI-StA (PLGA-PS) had an average diameter of~350 to 390 nm and a zeta potential of ~ -13 to -19 mV, respectively. The encapsulation efficiency of siRNA in PLGA-P and PLGA-PS was 26% and 43%, respectively. An uptake study by fluorescence microscopy suggested dendritic cells uptake and endosomal localization of both nanoparticles types. After exposure to B16F10 conditioned medium, dendritic cells showed high STAT3 and low CD86 expression indicating impaired function. STAT3 silencing by PLGA-P and PLGA-PS of STAT3 siRNA restored dendritic cells maturation and functionality as evidenced by the upregulation of CD86 expression, high secretion of TNF-α and significant allogenic T cell proliferation. Moreover, encapsulation in PLGA nanoparticles significantly reduced PEI-associated toxicity on dendritic cells.

Lee *et al.* conjugated PLGA to the 3' end of siRNA via a disulfide bond to synthesize hybrid conjugates. The PLGA-siRNA conjugates were spontaneously self-assembled to form a spherical core/shell type micellar structure of ~20 nm in an aqueous environment, probably by hydrophobic interaction of PLGA blocks in the core surrounded by siRNA shell layer. When LPEI was added to the PLGA-siRNA micelles in aqueous solution, stable micelles with a size of ~30 nm were produced via ionic complexation between siRNA and LPEI in the outer shell. The cationic PLGA-siRNA-LPEI micelles showed superior intracellular uptake and enhanced gene silencing effect, compared to naked LPEI-siRNA complexes. Naked siRNA formed unstable micron sized aggregates with LPEI (25 kDa) due to its low charge density and stiffness, resulting in low extent of cellular uptake. On the contrary, PLGA-siRNA conjugate micelles coated with both LPEI 25 kDa and 2.5 kDa at N/P ratio of

20 and 40, respectively, showed significantly higher intracellular uptake compared to the corresponding LPEI-siRNA complexes at the same N/P ratios. The improved cellular uptake was thought to be due to the stable PLGA-siRNA-LPEI conjugate micelles with a size of less than ~50 nm and highly cationic surface charge for facile cellular interactions. The LPEI 25 kDa-coated PLGA-siRNA conjugate micelles at the N/P ratio of 10, 20, and 40 downgraded the GFP expression to 85.2 ± 2.4, 57.1 ± 6.6, and $42.9 \pm 4.6\%$, respectively, while no significant changes in GFP silencing for the cells treated with naked LPEI 25 kDa-siRNA complexes at the same N/P ratios.

Recently, novel biodegradable monomethoxyPEG-PLGA-PLL (mPEG-PLGA-PLL) nanoparticles encapsulating siRNA against the production of Platelet-derived growth factor BB (PDGF-BB) from rat RPE cells were investigated to determine whether ultrasound (US) and/or US targeted microbubbles (MBs) can facilitate the delivery of nanoparticles loading siRNA to rat RPE cells. The mean diameter of mPEG-PLGA-PLL nanoparticles was 151 ± 74 nm and the zeta potential was -0.13 mV. The encapsulation efficiency of siRNA in mPEG-PLGA-PLL nanoparticles was 86.06%. The nanoparticles sustained the long-lasting siRNA release over one month after the initial burst of 20–30%. The data obtained showed that low intensity US or 15–20% MBs could increase the delivery efficiency of a lower concentration of mPEG-PLGA-PLL nanoparticles loading siRNA to RPE-J cells, whereas the combination of US with MBs under the optimal conditions for the enhancement of nanoparticles delivery did not further increase the cellular uptake of NPs compared to either US or MBs alone. Under the optimal condition for US-enhanced nanoparticles delivery, the enhanced PDGF-BB gene silencing with a combination of US and NPs encapsulating siRNA resulted in a significant decrease of mRNA and protein expression levels compared to nanoparticles alone. The study proposes that combination of the chemical (mPEG-PLGA-PLL nanoparticles loading siRNA) and physical (US) approaches could more effectively downregulate the mRNA and protein expression of PDGF-BB.

14.3 Future Perspectives

Rapid progress has been done in the development of non-condensing delivery systems during the last decade. More and more polymers have been investigated as non-condensing systems for siRNA delivery. Irrespective of both natural and synthetic polymers, safety, biodegradability and biocompatibility are the obligatory requirements. Although, PLGA-based nanoparticles have shown promising results with the encapsulation of nucleic acids, compared to viral vectors and cationic polymers, they still need to improve transfection efficiency and efficacy. The research on non-condensing gene therapy will advance with increased knowledge and innovative delivery strategies. With continuous development, PLGA-based siRNA therapeutics will eventually lead toward better treatments for many diseases.

14.4   Conclusions

PLGA-based nanoparticles have been successfully employed for *in vitro* and *in vivo* gene delivery studies and recently in siRNA delivery. Introduction of cationic polymers in PLGA nanoparticles significantly improved the encapsulation of siRNA and its release. Encapsulation of siRNA in cationic polymer modified PLGA nanoparticles has significantly improved the *in vitro* gene knockdown efficacy. Further, improvements in the stability of PLGA nanoparticles in the presence of serum along with lack of cytotoxicity favors the application of PLGA nanoparticles. However, the use of mixture of different polymers to achieve better encapsulation and security will pave the way towards implication of PLGA nanoparticles for siRNA therapeutics.

References

[1] Merdan, T., J. Kopecek, and T. Kissel. Prospects for cationic polymers in gene and oligonucleotide therapy against cancer. *Advanced Drug Delivery Reviews*, 54: 715–758, 2002.

[2] Kommareddy, S., and M. Amiji. Preparation and evaluation of thiol-modified gelatin nanoparticles for intracellular DNA delivery in response to glutathione. *Bioconjugate Chemistry*, 16: 1423–1432, 2005.

[3] Lemieux, P., N. Guerin, G. Paradis, R. Proulx, L. Chistyakova, A. Kabanov, and V. Alakhov. A combination of poloxamers increases gene expression of plasmid DNA in skeletal muscle. *Gene Therapy*, 7: 986–991, 2000.

[4] Morille, M., C. Passirani, A. Vonarbourg, A. Clavreul, and J.P. Benoit. Progress in developing cationic vectors for non-viral systemic gene therapy against cancer. *Biomaterials*, 29: 3477–3496, 2008.

[5] Bhavsar, M.D., and M.M. Amiji. Polymeric nano- and microparticle technologies for oral gene delivery. *Expert Opinion on Drug Delivery*, 4: 197–213, 2007.

[6] Panyam, J., and V. Labhasetwar. Biodegradable nanoparticles for drug and gene delivery to cells and tissue. *Advanced Drug Delivery Reviews*, 55: 329–347, 2003.

[7] Panyam, J., W.Z. Zhou, S. Prabha, S.K. Sahoo, and V. Labhasetwar. Rapid endo-lysosomal escape of poly(DL-lactide-co-glycolide) nanoparticles: implications for drug and gene delivery. *FASEB Journal*, 16: 1217–1226, 2002.

[8] Panyam, J., and V. Labhasetwar. Sustained cytoplasmic delivery of drugs with intracellular receptors using biodegradable nanoparticles. *Molecular Pharmaceutics*, 1: 77–84, 2004.

[9] Katas, H., E. Cevher, and H.O. Alpar. Preparation of polyethyleneimine incorporated poly(d,l-lactide-co-glycolide) nanoparticles by spontaneous emulsion diffusion method for small interfering RNA delivery. *International Journal of Pharmaceutics*, 369: 144–154, 2009.

[10] Tahara, K., H. Yamamoto, N. Hirashima, and Y. Kawashima. Chitosan-modified poly(D,L-lactide-co-glycolide) nanospheres for improving siRNA delivery and gene-silencing effects. *European Journal of Pharmaceutics and Biopharmaceutics*, 74: 421–426, 2010.

[11] Yuan, X., B. Shah, N. Kotadia, J. Li, H. Gu, and Z. Wu. The development and mechanism studies of cationic chitosan-modified biodegradable PLGA nanoparticles for efficient siRNA drug delivery. *Pharmaceutical Research*, 27: 1285–1295, 2010.

[12] Zeng, P., Y. Xu, C. Zeng, H. Ren, and M. Peng. Chitosan-modified poly(d,l-lactide-co-glycolide) nanospheres for plasmid DNA delivery and HBV gene-silencing. *International Journal of Pharmaceutics*, 415: 259–266, 2011.

[13] Nguyen, J., T.W.J. Steele, O. Merkel, R. Reul, and T. Kissel. Fast degrading polyesters as siRNA nano-carriers for pulmonary gene therapy. *Journal of Controlled Release*, 132: 243–251, 2008.

[14] Csaba, N., P. Caamano, A. Sanchez, F. Dominguez, and M.J. Alonso. PLGA:poloxamer and PLGA:poloxamine blend nanoparticles: new carriers for gene delivery. *Biomacromolecules*, 6: 271–278, 2005.

[15] Csaba, N., A. Sanchez, and M.J. Alonso. PLGA:poloxamer and PLGA:poloxamine blend nanostructures as carriers for nasal gene delivery. *Journal of Controlled Release*, 113: 164–172, 2006.

[16] Luo, G., C. Jin, J. Long, D. Fu, F. Yang, J. Xu, X. Yu, W. Chen, and Q. Ni. RNA interference of MBD1 in BxPC-3 human pancreatic cancer cells delivered by PLGA-poloxamer nanoparticles. *Cancer Biology & Therapy*, 8: 594–598, 2009.

[17] Patil, Y., and J. Panyam. Polymeric nanoparticles for siRNA delivery and gene silencing. *International Journal of Pharmaceutics*, 367: 195–203, 2009.

[18] Andersen, M.Ø., A. Lichawska, A. Arpanaei, S.M. Rask Jensen, H. Kaur, D. Oupicky, F. Besenbacher, P. Kingshott, J. Kjems, and K.A. Howard. Surface functionalisation of PLGA nanoparticles for gene silencing. *Biomaterials*, 31: 5671–5677, 2010.

[19] Wang, J., S.-S. Feng, S. Wang and Z.-y. Chen. Evaluation of cationic nanoparticles of biodegradable copolymers as siRNA delivery system for hepatitis B treatment. *International Journal of Pharmaceutics*, 400: 194–200, 2010.

[20] Cun, D., C. Foged, M. Yang, S. Frøkjær, and H.M. Nielsen. Preparation and characterization of poly(dl-lactide-co-glycolide) nanoparticles for siRNA delivery. *International Journal of Pharmaceutics*, 390: 70–75, 2010.

[21] Cun, D., D.K. Jensen, M.J. Maltesen, M. Bunker, P. Whiteside, D. Scurr, C. Foged, and H.M. Nielsen. High loading efficiency and sustained release of siRNA encapsulated

in PLGA nanoparticles: Quality by design optimization and characterization. *European Journal of Pharmaceutics and Biopharmaceutics*, 77: 26–35, 2011.

[22] Alshamsan, A., A. Haddadi, S. Hamdy, J. Samuel, A.O.S. El-Kadi, H. Uludag? and A. Lavasanifar. STAT3 Silencing in dendritic cells by siRNA polyplexes encapsulated in PLGA nanoparticles for the modulation of anticancer immune response. *Molecular Pharmaceutics*, 7: 1643–1654, 2010.

[23] Lee, S.H., H. Mok, Y. Lee and T.G. Park. Self-assembled siRNA-PLGA conjugate micelles for gene silencing. *Journal of Controlled Release*, 152: 152–158, 2011.

[24] Du, J., Q.S. Shi, Y. Sun, P.F. Liu, M.J. Zhu, L.F. Du and Y.R. Duan. Enhanced delivery of monomethoxypoly(ethylene glycol)-poly(lactic-co-glycolic acid)-poly l-lysine nanoparticles loading platelet-derived growth factor BB small interfering RNA by ultrasound and/or microbubbles to rat retinal pigment epithelium cells. *Journal of Gene Medicine*, 13: 312–323, 2011.

15

Nanomaterials for siRNA Delivery

15.1   Introduction

Progress in the fabrication of inorganic materials has led to a plethora of applications in the field of biomedical therapeutics. The development of robust and versatile chemical syntheses has provided tools for engineering inorganic nanomaterials with different sizes, shapes, and other properties. Studies reported that nanoparticles with different sizes and shapes exhibit differential uptake by cells, e.g., 50 nm spherical nanoparticles were preferentially taken up over 14, 30, 74, and 100 nm nanoparticles, and also over nanorods with different aspect ratios. Moreover, inorganic nanomaterials are also amenable to various strategies for incorporation of desired functionalities. The surface properties of nanoparticles can also be fabricated to desired thickness and composition with polymers, biomolecules, inorganic coatings, and also smaller preformed nanoparticles. Nanomaterials such as gold nanoparticles, carbon nanotubes, and silica nanoparticles attracted considerable attention in their pharmaceutical applications due to their biocompatibility and ability to facilitate the delivery of therapeutic cargos. Also, coating of nanomaterials with cationic polymers has found numerous applications in gene and siRNA delivery.

15.2   Gold Nanoparticles

AuNPs have attracted increasing attention in a wide variety of therapeutic areas including DNA mismatch detection, biomolecular sensing, and hyperthermal cancer therapy Recently, several attempts have been made to utilize AuNPs for

S. Nimesh and R. Chandra,
Theory, Techniques and Applications of Nanotechnology in Gene Silencing, 177–192.
© 2011 *River Publishers. All rights reserved.*

(a)   (b)

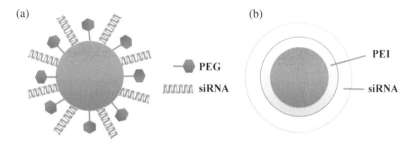

Fig. 15.1 Schematic representation of AuNPs (a) siRNA coupled to AuNPs via thiol bonding, (b) siRNA adsorbed onto AuNPs coated with cationic polymer such as PEI.

gene delivery due to their low toxicity and ease of surface functionalization that in turn allows for selective tuning of surface properties such as charge and hydrophobicity. AuNPs chemically modified with primary and quaternary amine moieties used to deliver plasmid DNA, exhibited more efficient intracellular delivery than the conventional transfection agents. Thiol (SH)-modified antisense oligonucleotides were directly conjugated onto AuNPs for regulation of protein expression in cells. Additionally, AuNPs were suggested to be functionalized with various antibodies, peptides, and small molecular ligands to achieve a target cell-specific uptake, endosome escape, and nuclear localization for efficient gene delivery. The strategies developed for functionalization can be classified as: (i) siRNA coupled directly to the AuNPs surface via a gold–thiol bond, and (ii) siRNA adsorbed onto the AuNPs surface though electrostatic interactions. Both strategies have been found to incorporate PEG or other shielding polymer in one form or another to help stabilize and/or promote endosomal escape into the cytoplasm (Figure 15.1).

15.2.1 Conjugates Formed via Thiol Chemistry

The gold–thiol interaction has been employed since decades to immobilize thiol-containing macromolecules to gold surfaces. This bonding is quite dynamic which disassociates under reducing conditions such as the high glutathione concentrations in the cell cytoplasm. Its easily coupling and dynamic nature are well exploited for fabrication of siRNA-AuNPs conjugates. The first study on siRNA-AuNPs reports co-decoration of 15 nm AuNPs with SH-PEG$_{5000}$-PAMA$_{7500}$ and SH-siRNA. PAMA is a poly (2-N, N-dimethyamino)ethyl methacrylate polymer bearing terminal amines. The authors reported 45 siRNA molecules per AuNP, in agreement with earlier literature reporting DNA loading on AuNPs. With a 100 nM siRNA dose after 24 h incubation, a 65% knockdown of luciferase expression in HuH-7 cells was achieved. In this complex, the siRNA ($\sim$14 kDa) is roughly the same size as the co-loaded PEG-PAMA ($\sim$12.5 kDa). Later, a variation on this concept was reported where 13 nm AuNPs were co-loaded with SH-PEG$_{400}$ and SH-siRNA. In this system, the PEG was significantly smaller than the siRNA such that the particle surface presented to the cells displays almost entirely siRNA. The authors report a similar siRNA loading of 33 duplexes per particle. With 100 nM siRNA dosing after four days of incubation, 70% luciferase knockdown was observed in HeLa cells. Confocal microscopy imaging shows a high level of Cy5-labeled conjugates inside cells, although colocalization staining with an endosomal/lysosomal stain was not reported.

These siRNA-AuNPs are interesting with respect to the high uptake efficiencies (nearly 100%) in cells. The process of nanoparticles internalization by HeLa cells was determined to be mediated by scavenger receptors, which is a group of structurally unrelated molecules known to mediate the endocytosis of certain polyanionic ligands, including nucleic acids. The molecular interaction is postulated to be based on these receptors forming a quadraplex like structure with nucleic acid on the AuNPs surface. As more and more siRNA is loaded on the AuNPs surface, this molecular interaction is increased, thereby leading to enhanced internalization. Further, studies in a mouse cell line showed similar results proposing that this pathway may be conserved across some species.

Activation of immune system potentiated by stimulation of TLR 7 and 8 is one of the most crucial barriers toward siRNA therapy. Further studies with siRNA-AuNPs particles have reported a reduced response of the innate immune system compared to "interferon stimulatory" DNA (ISD) delivered via complexation with Lipofectamine. After a 4 h incubation period, ISD-AuNPs were reported to induce a 25-fold less activation of IFN-β as determined by qRT-PCR. IFN-β is a type of interferon, proteins produced by the immune system in response to pathogens. IFN-β levels were found to vary inversely proportional to the density of the DNA on the AuNP surface, suggesting that tight packing of the DNA may prohibit interaction with cellular DNA binding proteins. The IFN-β response to siRNA-AuNPs was reported to be about 1/3 of that compared to siRNA delivered via Lipofectamine 2000.

While most of the siRNA-AuNPs reported employed a co-loaded PEG-siRNA particle, a different strategy was adopted while approaching of the first loading 15 nm AuNPs surface with a SH-PEG$_{1000}$-NH$_2$ and then attaching siRNA using a disulfide crosslinker, N-succinimidyl 3-[2-pyridyldithio]-propionate (SPDP), to the terminal amine on the PEG. This forms AuNP with a corona of siRNA attached via a cleavable disulfide linkage. A comparable siRNA loading of 40 duplexes per particle has been reported. These particles were then coated with one of 14 different PBAEs to enhance the cellular uptake and further endosomal escape. PBAEs are a new class of cationic biodegradable polymer that has shown great promise for plasmid delivery in various cell lines. With a 120 nM siRNA dose after a 24 h incubation, luciferase expression in HeLa cells was decreased by >90%. In this case the siRNA-AuNPs without PBAEs did not show any silencing.

To address the issues related to the endosomal escape, 40 nm hollow gold nanoshells (AuNSs) were employed. These materials are susceptible to local heating when irradiated with NIR light which, at low powers, is not destructive to tissue. In the first example, AuNSs are co-loaded with SH-siRNA-PEG$_{300}$ and SH-PEG$_{3000}$. Approximately 10^2–10^3 siRNA were loaded on AuNS. The conjugates were then coated with a TAT lipid for uptake into cells. TAT is a cell internalizing peptide derived from HIV-1 trans-activator peptide. The lipid-coated complexes were then added to C166 mouse endothelial cells expressing GFP and irradiated ($3.5\,\mathrm{W\,cm^{-2}}$, 30 s) the following day. Imaging of GFP was performed over the next three days. By day two, the GFP expression was reduced by 80% normalized to control cells treated with non-targeting complexes and without radiation. Samples that were treated with siRNA complexes which were not irradiated or irradiated with less power did not show silencing. Confocal microscopy of Cy3-siRNA labeled complexes show decomplexation when irradiated with lower power ($2.4\,\mathrm{W\,cm^{-2}}$, 2 s) but siRNA was maintained within the endosomal housing unless irradiated with the higher setting.

In another study, 40 nm AuNSs used for delivering siRNA were reported to exhibit *in vivo* gene knockdown. In these complexes, AuNSs were co-loaded with SH-siRNA and TA-PEG$_{5000}$-F (TA-thioctic acid, F-folic acid). The thioctic acid allowed attachment to the gold surface and the folic acid provided targeting capabilities. Conjugates were delivered to HeLa cells cultured in reduced folate media at a particle concentration of 0.5 nM (siRNA loading was not reported). Interestingly, the authors report no cellular uptake with particles lacking the folate-targeting group. Particles prepared with the folate-targeting group showed cellular uptake after 2 h. However, particles prepared with dye-labeled Dy547-siRNA showed colocalization in cells that were also stained with LysoTracker Green, indicating that the particles were trapped in the endosomal compartments after 2 h. Endosomal escape was reported 15 min post-irradiation (800 nm, 50 mW cm^{-2}, 60 s). Following photothermal transfection, p65 silencing was measured via western blot showing silencing of 30%, 92%, 95%, and 84% at time points of 24, 48, 72, and 96 h. The authors report no silencing without the targeting folate or without irradiation. Further, *in vivo* experiments were performed administering these conjugates via intravenous injection into nude mice bearing subcutaneous HeLa cervical cancer xenografts. Following injection of folate-bearing conjugates, p65 was downregulated to 23% of that from the contralateral tumors in the same mouse not irradiated with NIR laser. Suppression of p65 showed increased sensitivity to chemotherapeutic treatment with irinotecan. It should be noted that off-target immune responses have been activated when non-chemically modified siRNA, similar to that employed in this study, has been used.

15.2.2 Complexes Formed via Electrostatic Interactions

An alternative approach to direct gold–thiol conjugation is the electrostatic interaction between the negatively charged siRNA molecules and the AuNPs surface. The most common strategy involves citrate reduction in aqueous solution to form AuNPs. Various layer-by-layer methodologies took advantage of the negative charge of these materials and incorporated cationic polymer PEI. In one study, 15 nm AuNP core followed by layers of PEI and siRNA were prepared. In this system, some particles have a terminal siRNA layer while an alternative particle has a terminal PEI layer. Both particle types are reported to have a loading density of 780 siRNA/AuNP. The final size of these particles is reported to be between 20–25 nm. While this size is slightly larger than the 15 nm AuNP and greater siRNA loading would be expected accordingly, the loading density is significantly higher than that found for thiol–gold conjugated particles or other particles. This may be due to a strong electrostatic interaction between PEI and siRNA. The cellular uptake of these particles was examined in GFP-expressing CHO-K1 cells after 6 h. Significantly, higher particle uptake was reported when the final layer was composed of siRNA; however, TEM images of fixed cells show these particles primarily trapped within the endosome. GFP silencing studies report 70% GFP knockdown 48 h post-transfection with particles containing an external PEI layer and no knockdown for particles with a terminal siRNA layer (0.37 nM AuNP, ~288 nM siRNA). This suggests a need for an endosomal escape mechanism using particles of this type since significant particle uptake was seen without PEI but no knockdown was observed.

A different layer-by-layer study reports a similar method that incorporates an anionic charge-reversal polyelectrolyte (PAH-Cit = cis-aconitic anhydride-functionalized poly(allylamine)) with PEI and siRNA AuNP complex. In this system, the PAH-Cit undergoes a charge reversal during a pH change from 7.4 to 5.0. The intent was to cause the complex to disassemble and release the siRNA during the acidification of the endosome. Lamin A/C protein expression, normalized against the protein GAPDH, was examined by western blot in HeLa cells after 48 h incubation with a 1 μg dose (in 1 well of a six-well plate) of siRNA in complex form. Silencing of 80% was reported, compared to only 20% silencing seen for complexes formed with non-charge-reversal polymer used in place of PAH-Cit, suggesting a functional role of the charge-reversal polymer. Confocal images were taken 6 h after incubation with Cy5-labeled complexes and showed increased endosomal escape compared to complexes made without PAH-Cit. An alternative layering method incorporates PEI_{25000} as the reducing and stabilizing agent during the synthesis of 15 nm AuNPs in place of citrate. This results in AuNP with a positive surface charge which is layered with siRNA molecules via electrostatic interactions. MDA-MB-435 cells incubated with these complexes at 80 nM siRNA concentrations were reported to have an 80% reduction in GFP expression after 48 h.

In addition to the polymer-coated particles, AuNPs have also been synthesized in the presence of cysteamine hydrochloride to affect amine-functionalized AuNPs. siRNA-PEG_{5000} was added to these particles, forming a complex with siRNA electrostatically adhered to the surface with a dangling PEG tail. The siRNA was attached to the PEG via a cleavable disulfide bond allowing disassociation of the PEG in the reductive cellular environment. While the AuNPs themselves were 15 nm, addition of siRNA-PEG_{5000} produced particles of 96 nm. These particles were composed of clusters of ~10 AuNPs but the final siRNA content was not determined. GFP expressing PC-3 cells were incubated with 60 nM siRNA in complexed form to affect a GFP silencing of 80%. Complexes formed with siRNA (no PEG) were not stable and formed micrometer-sized aggregates within 15 min and were not readily endocytosed by the cells.

15.3 Mesoporous Silica Nanoparticles

Silica-based nanoparticles have been investigated as nucleic acids delivery vectors due to their inherent biocompatibility and low toxicity. Moreover, the silanol surface chemistry of silica nanoparticles provides ease of modification and functionalization for enhanced nucleic acids delivery (Figure 15.2). Unique structural and functional properties of MSNs such as chemically stable mesoporous structures, large surface areas, tunable pore sizes, encapsulation of small molecules, and well-defined surface chemistry made them fascinating delivery system. The first study of MSNs-mediated siRNA delivery involved the incorporation of PEI of various MW ranging from 0.6 to 25 kDa, to the surface of MSN through non-covalent coupling. The nanoparticles were in the size range of 100–130 nm with a uniform pore size of ~2.5 nm, as shown by TEM. All of the non PEI-coated particles exhibited a negative zeta potential, whereas PEI-coated particles showed a positive charge. The attachment of PEI not only increased MSNs cellular uptake but also generated a cationic surface to which DNA and siRNA constructs could be attached. Cellular uptake of

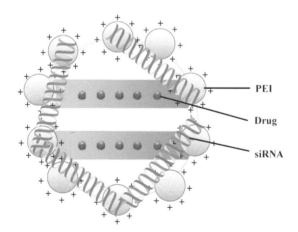

Fig. 15.2 Mesoporous silica nanoparticle coated with cationic polymer such as PEI for co-delivery of antitumor drug and siRNA.

FITC-labeled MSNs was assessed by flow cytometry and confocal microscopy in PANC-1, BxPC3, RAW 264.7, and BEAS-2B cells. The uptake of MSNs-PEI-25 kDa and MSNs-PEI-1.2 kDa was two-fold higher when compared with the phosphonate or PEG-coated particles. The particles coated with the 10 kDa PEI polymer were particularly efficient for transducing HEPA-1 cells with a siRNA construct that was capable of knocking down GFP expression. The MSN-PEI/siRNAs coated with the 10 kDa and 25 kDa PEI were capable of knocking down GFP level by 55% and 60%, respectively, in GFP expressing HEPA-1 cells. Further, the enhanced cellular uptake of the non-toxic cationic MSNs enhanced the delivery of the hydrophobic anticancer drug, paclitaxel, to pancreatic cancer cells.

To investigate the utility of MSNs for cancer therapy, MSNs were employed to codeliver doxorubicin (Dox) (as a model apoptosis-inducing anticancer drug), and siRNA (as a suppressor of cellular anti-apoptotic defence) simultaneously into multidrug-resistant cancer cells. The Dox-loaded MSNs were modified with G2 amine-terminated PAMAM dendrimer, which can complex with siRNAs targeted against mRNA encoding Bcl-2 protein. The morphology of MSN-Dox-G2 complex with siRNA was visualized by TEM, showed that a $\sim$20 nm thick layer of siRNAs was formed surrounding the surface of MSN-Dox-G2. The cellular internalization and intracellular release of Dox and siRNA studies in A2780/AD human ovarian cancer cells, suggested of efficient co-delivery of siRNA into cytoplasm. In the cells incubated with MSN-Dox-G2/siRNA complexes for 24 h, the Bcl-2 mRNA level was effectively suppressed by $\sim$80 % while cells incubated with MSN-Dox-G2 without Bcl-2 siRNA showed similar Bcl-2 mRNA level as the control cells without treatment. Moreover, the anticancer efficacy of Dox increased 132 times compared to free Dox, owing to the simultaneous delivery of siRNA.

In a similar study, MSNs derivatized with PEI used to codeliver Dox and Pgp siRNA to a drug-resistant cancer cell line (KB-V1 cells) to accomplish cell killing in synergistic fashion. The size of the nanoparticles prepared by binding of PEI in the size range 1.8–25 kDa to the phosphonate-MSNs surface was 100–120 nm, exhibited uniform pore sizes of 2–2.5 nm. The siRNA binding capacity of nanoparticles increased with increase in size of PEI polymers, indicating that all siRNA was bound at a N/P ratio > 16 (PEI 1.8 kD), > 16 (PEI 10 kD) and > 8 (PEI 25 kD). The assessment of the efficacy of Pgp siRNA delivered by nanoparticles as estimated by western blotting; demonstrated 80 or 90% reduction in MDR-1 expression in cells treated with MSNs-coated particles that contain 10 and 25 kDa PEI. These studies clearly demonstrate the possible use of MSNs platform to effectively deliver a siRNA that knocks down gene expression of a drug exporter that can be used to improve drug sensitivity to a chemotherapeutic agent.

In order to effectively shutdown both exogenous and endogenous gene signals, MSNs modified with PEI were used to deliver siRNA to mammalian cells. The ability of PEI-MSNs to complex with siRNA was confirmed by UV absorption measurements and gel assay which suggested of negligible release of siRNA. The exogenous gene silencing efficacy of MSNs loaded with siRNA against EGFP was tested in PANC-1 cells stably expressing EGFP. The MSNs showed comparable or even better efficacy of delivering siRNA in silencing EGFP ($\sim$ 61.7% downregulation) compared with Lipofectamine. Furthermore, the endogenous gene silencing efficiency of MSNs was assessed by delivering siRNA against Akt in PANC-1cells, significantly reduced Akt expression. Also, the nanoparticles effectively delivered siRNA against K-ras, which is another critical target for cancer therapeutics.

In a recent study, the surface of MSNs was modified with PEG and poly (2-(dimethylamino)ethylmethacrylate) (PDMAEMA) or poly (2-(diethylamino) ethylmethacrylate) (PDEAEMA). The average size of the unmodified MSNs was 56.7 nm by TEM, which increased to $\sim$100nm after PEG coating followed by further increase in size to $\sim$130nm and $\sim$170 in the case of PDMAEMA and PDEAEMA respectively. The ability of the polycation-coated MSNs to deliver siRNA was evaluated using only PDMAEMA conjugated MSNs because of their superior performance in delivering DNA. Luciferase expression decreased to $\sim$50% of the control non-transfected cells when antiLuc siRNA was used, while no significant silencing was observed with control non-targeted siRNA.

15.4 Carbon Nanotubes

Carbon nanotubes (CNTs) are well defined, hollow cylindrical graphene nanomaterials with very high aspect ratios, lengths from several hundred nanometers to several micrometers and diameter of 0.4–2 nm for SWNTs to 2–100 nm for MWNTs. The structure of SWNTs can be considered as graphene sheet rolled up into a seamless hollow cylinder and MWNTs visualized as several

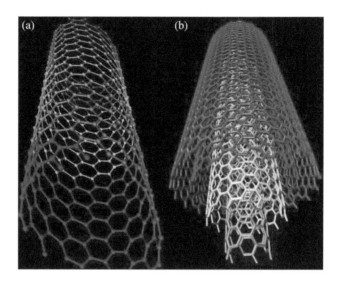

Fig. 15.3 Representative structures of carbon nanotubes, (a) SWNT and (b) MWNT.

co-axially arranged SWNTs of different radii with an inter-tube separation close to the inter-plane separation in graphite (0.34–0.35 nm) (Figure 15.3). CNTs have attracted much attention as a platform for the delivery of drugs and biomolecules due to their variety of size- and structure-dependent optical, thermal, electrical, and mechanical properties, as well as their unique ability to traverse across the cell membrane. Although nanotubes are insoluble in water, they can be derivatized to render them water soluble and biocompatible for a range of biological applications. One of the first studies involved functionalization of short SWNTs by non-covalent adsorption of phospholipid molecules with PEG (2 kDa) (PL-PEG) chains and terminal amine or maleimide groups (PL-PEG-NH2 or PL-PEG-maleimide). The disulfide bond was incorporated by employing a heterobifunctional crosslinker sulfosuccinimidyl 6-(3'-[2-pyridyldithio]propionamido)hexanoate (sulfo-LC-SPDP) for any thiol-containing biomolecule (X) to form SWNT-PL-PEG-SS-X (where X can be DNA or siRNA). These SWNTs were employed to deliver siRNA known to silence the gene encoding lamin A/C protein present inside the nuclear lamina of cells. Confocal imaging revealed significant reduction in lamin A/C protein expression by siRNA relative to untreated control cells.

The potentiality of SWNTs was investigated to deliver siRNA to reach, enter and genetically modify dendritic cells *in vivo*. Positively charged SWNTs were prepared by using 1,6-diaminohexane as confirmed by TEM and AFM. The functionalized SWNTs absorbed the siRNA to form complexes. These complexes were preferentially taken up by splenic CD11c + DCs, CD11b + cells and also Gr-1 + CD11b + cells comprising dendritic cells, macrophages and other myeloid cells to silence the targeting gene. Suppressor of cytokine signaling 1 (SOCS1) restricts the ability of dendritic cells to break self-tolerance and induce anti-tumor immunity. Infusion of SWNTs carrying SOCS1 siRNA reduced SOCS1 expression and retarded the growth of established B16 tumor in mice, indicating the possibility of *in vivo* immunotherapeutics using SWNTs-based siRNA transfer system. In another approach, Zhang *et al.* used ammonium functionalized SWNTs to conjugate siRNA and examined the ability of telomerase reverse transcriptase (TERT) siRNA delivered via SWNT complexes to silence the expression of TERT and inhibit the proliferation and growth of tumor cells both *in vitro* and in mouse models. The functionalized SWNTs -$CONH$-$(CH_2)_6$-$NH_3^+Cl^-$ facilitated the coupling of siRNAs to form the mTERT siRNA:SWNT+ complex. These SWNTs rapidly entered three cultured murine tumor cell lines, suppressed mTERT expression, and produced growth arrest. Injection of mTERTsiRNA:SWNT+ complexes into subcutaneous Lewis lung tumors reduced tumor growth. Furthermore, human TERT siRNA:SWNT+ complexes also suppressed the growth of human HeLa cells both *in vitro* and when injected into tumors in nude mice. The ammonium-functionalized SWNTs was also employed to deliver siRNA targeted to cyclin A2 in chronic myelogenous leukemia K562 cells, resulting in suppression of cyclin A2 expression. The depletion of cyclin A2 causes cell proliferation arrest and promotes apoptosis of chronic myelogenous leukemia K562 cells.

In another strategy, SWNTs were functionalized with hexamethylenediamine (HMDA) and poly(diallyldimethylammonium) chloride (PDDA) to render positive charge to bind siRNA. The PDDA-HMDA-SWNTs exhibited negligible cytotoxic effects on isolated rat heart cells at concentrations up to 10 mg/l. PDDA-HMDA-SWNTs loaded with extracellular signal-regulated kinase (ERK) siRNA were able to cross the cell membrane and suppressed expression of the ERK target proteins in primary cardiomyocytes by about 75%. Further, amino functionalized MWNTs were used for the treatment of a human lung carcinoma model *in vivo* using siRNA. The MWNT-NH_3^+:siRNA complexes administered intratumorally elicited delayed tumor growth and increased survival of xenograft-bearing animals. Moreover, siTOX delivery via the cationic MWNT-NH$_3^+$ was biologically active *in vivo* by triggering an apoptotic cascade, leading to extensive necrosis of the human tumor mass. Furthermore, a new approach was described for delivering siRNA into cancer cells by non-covalently complexing unmodified siRNA with pristine single walled carbon nanotubes (SWCNTs). The complexes were prepared by a simple sonication of pristine SWCNTs in a solution of siRNA, which then served both as the cargo and as the suspending agent for the SWCNTs. When complexes containing siRNA targeted to hypoxia-inducible factor 1 alpha (HIF-1α) were added to cells growing in serum containing culture media, there was strong specific inhibition of cellular HIF-1α activity. The ability to obtain a biological response to SWCNT/siRNA complexes was seen in a wide variety of cancer cell types. Moreover, intratumoral administration of SWCNT-HIF-1α siRNA complexes in mice bearing MiaPaCa-2/HRE tumors significantly inhibited the activity of tumor HIF-1α.

A novel approach for siRNA cellular delivery using siRNA coiling into carboxyl-functionalized SWCNTs was described. The SWCNT-siRNA delivery system demonstrated non-specific toxicity and transfection efficiency greater than 95%. This approach offers the potential for siRNA delivery into different types of cells, including hard-to-transfect cells, such as neuronal cells and cardiomyocytes. The SWCNT-siRNA system was also tested in a non-metastatic human hepatocellular carcinoma cell line (SKHep1). In all types of cells used, the SWCNT-siRNA delivery system showed high efficiency and apparent no side effects for various *in vitro* applications.

Caspase-3 activation contributes to brain tissue loss and downstream biochemical events that lead to programmed cell death after traumatic brain injury. Al-Jamal *et al.* investigated the effect of silencing of Caspase-3 using multiple walled carbon nanotubes (MWCNTs)-mediated *in vivo* RNAi. Effective delivery of siRNA directly to the CNS has been shown to normalize phenotypes in animal models of several neurological diseases. It was shown that perilesional stereotactic administration of a Caspase-3 siRNA delivered by functionalized MWCNTs reduced neurodegeneration and promoted functional preservation before and after focal ischemic damage of the rodent motor cortex using an endothelin-1 induced stroke model.

In a recent study, MWCNTs covalently modified with low MW PEI showed that the resulting nanohybrids were able to bind siRNA, enter cells, and induce endosomal escape. The PEI-MWCNTs siRNA delivery system showed superior performances compared to a reference lipidic carrier (jet-siENDO), even in the presence of seric proteins. Along a similar line, Varkouhi *et al.* investigated PEI and pyridinium functionalized MWCNTs for siRNA delivery. Both functionalized MWCNTs complexed siRNA and showed 10–30% silencing activity and a cytotoxicity of 10–60%. However, in terms of reduced toxicity or increased silencing activity, MWCNT-PEI and MWCNT-pyridinium did not show an added value over PEI and other standard transfection systems.

15.5 Future Perspectives

Gene therapy, especially the recent progress in siRNA brings great hopes to cure some untreatable diseases. However, the foremost issue in gene silencing is to deliver these therapeutic nucleic acids to the targeted sites without eliciting toxicity. Nanomaterials such as carbon nanotubes, due to their large surface area, needle like shape, and a series of amazing electronic and optical properties, are expected to solve the aforementioned problems and develop a revolutionary delivery vehicle for gene therapy. Although it has been demonstrated that siRNA-AuNPs are shielded to some extent from activating the immune system, at some point, the siRNA detaches from the AuNP surface and may not continue to enjoy the shielding effects of the particle complex. Future *in vivo* work will need to carefully consider the correct choice of chemical modifications to incorporate into the siRNA molecules to avoid activation of an off-target immune response. A key remaining question is siRNA-nanomaterials biocompatibility, and in particular the potential for tissue accumulation. Additional biocompatibility studies will be needed for nanomaterials that deliver intracellular payloads. The most recent reports on the toxicity of MSNs *in vivo* have shown promising results. However, much more work needs to be performed to demonstrate the biocompatibility of MSNs *in vivo*. In addition, due to the versatility to further functionalize inorganic nanomaterials, it is believed that better systems can be developed to overcome any toxicological issue.

15.6 Conclusions

These studies on nanomaterials report increased siRNA stability due to the compact nature of the particles, reducing the ability for nucleases to degrade the siRNA. While colloidal gold has been used for the treatment of rheumatoid arthritis in humans, it is unclear how biocompatible AuNPs will be following intraveneous injection, in particular when the ultimate destination is the cytoplasm inside cells. However, an ongoing clinical trial has observed no toxicity to date with their AuNP formulations following intravenous injections. In the case of the MSNs-based target-specific nanovehicles, encouraging results have been obtained *in vitro*, but these systems have to be investigated *in vivo*. The nanohybrid vehicle appears very promising and more efficient systems could be envisioned using labile carbon nanotubes connecting bonds.

References

[1] Chithrani, B.D., A.A. Ghazani, and W.C. Chan. Determining the size and shape dependence of gold nanoparticle uptake into mammalian cells. *Nano Letters*, 6: 662–668, 2006.

[2] Hirsch, L.R., R.J. Stafford, J.A. Bankson, S.R. Sershen, B. Rivera, R.E. Price, J.D. Hazle, N.J. Halas, and J.L. West. Nanoshell-mediated near-infrared thermal therapy of tumors under magnetic resonance guidance. *Proceedings of the National Academy of Sciences of the United States of America*, 100: 13549–13554, 2003.

[3] Otsuka, H., Y. Akiyama, Y. Nagasaki, and K. Kataoka. Quantitative and reversible lectin-induced association of gold nanoparticles modified with alpha-lactosyl-omega-mercapto-poly(ethylene glycol). *Journal of the American Chemical Society*, 123: 8226–8230, 2001.

[4] Sandhu, K.K., C.M. McIntosh, J.M. Simard, S.W. Smith, and V.M. Rotello. Gold nanoparticle-mediated transfection of mammalian cells. *Bioconjugate Chemistry*, 13: 3–6, 2002.

[5] Niidome, T., K. Nakashima, H. Takahashi, and Y. Niidome. Preparation of primary amine-modified gold nanoparticles and their transfection ability into cultivated cells. *Chemical Communications (Cambridge)*, 7(17):1978–1979, 2004.

[6] Rosi, N.L., D.A. Giljohann, C.S. Thaxton, A.K. Lytton-Jean, M.S. Han, and C.A. Mirkin. Oligonucleotide-modified gold nanoparticles for intracellular gene regulation. *Science*, 312: 1027–1030, 2006.

[7] Hurst, S.J., A.K. Lytton-Jean, and C.A. Mirkin. Maximizing DNA loading on a range of gold nanoparticle sizes. *Analytical Chemistry*, 78: 8313–8318, 2006.

[8] Giljohann, D.A., D.S. Seferos, A.E. Prigodich, P.C. Patel, and C.A. Mirkin. Gene regulation with polyvalent siRNA–nanoparticle conjugates. *Journal of the American Chemical Society*, 131: 2072–2073, 2009.

[9] Patel, P.C., D.A. Giljohann, W.L. Daniel, D. Zheng, A.E. Prigodich, and C.A. Mirkin. Scavenger receptors mediate cellular uptake of polyvalent oligonucleotide-functionalized gold nanoparticles. *Bioconjugate Chemistry*, 21: 2250–2256, 2010.

[10] Massich, M.D., D.A. Giljohann, D.S. Seferos, L.E. Ludlow, C.M. Horvath, and C.A. Mirkin. Regulating immune response using polyvalent nucleic acid-gold nanoparticle conjugates. *Molecular Pharmaceutics*, 6: 1934–1940, 2009.

[11] Lee, J.-S., J.J. Green, K.T. Love, J. Sunshine, R. Langer, and D.G. Anderson. Gold, poly(β-amino ester) nanoparticles for small interfering RNA delivery. *Nano Letters*, 9: 2402–2406, 2009.

[12] Braun, G.B., A. Pallaoro, G. Wu, D. Missirlis, J.A. Zasadzinski, M. Tirrell. and N.O. Reich. Laser-activated gene silencing via gold nanoshell-siRNA conjugates. *ACS Nano*, 3: 2007–2015, 2009.

[13] Lu, W., G. Zhang, R. Zhang, L.G. Flores, 2nd, Q. Huang, J.G. Gelovani, and C. Li. Tumor site-specific silencing of NF-kappaB p65 by targeted hollow gold nanosphere-mediated photothermal transfection. *Cancer Research*, 70: 3177–3188, 2010.

[14] Marques, J.T., and B.R. Williams. Activation of the mammalian immune system by siRNAs. *Nature Biotechnology*, 23: 1399–1405, 2005.

[15] Elbakry, A., A. Zaky, R. Liebl, R. Rachel, A. Goepferich, and M. Breunig. Layer-by-layer assembled gold nanoparticles for siRNA delivery. *Nano Letters*, 9: 2059–2064, 2009.

[16] Guo, S., Y. Huang, Q. Jiang, Y. Sun, L. Deng, Z. Liang, Q. Du, J. Xing, Y. Zhao, P.C. Wang, A. Dong, and X.J. Liang. Enhanced gene delivery and siRNA silencing by gold nanoparticles coated with charge-reversal polyelectrolyte. *ACS Nano*, 4: 5505–5511, 2010.

[17] Song, W.-J., J.-Z. Du, T.-M. Sun, P.-Z. Zhang, and J. Wang. Gold nanoparticles capped with polyethyleneimine for enhanced siRNA delivery. *Small*, 6: 239–246, 2010.

[18] Lee, S.H., K.H. Bae, S.H. Kim, K.R. Lee, and T.G. Park. Amine-functionalized gold nanoparticles as non-cytotoxic and efficient intracellular siRNA delivery carriers. *International Journal of Pharmaceutics*, 364: 94–101, 2008.

[19] Xia, T. M. Kovochich, M. Liong, H. Meng, S. Kabehie, S. George, J.I. Zink, and A.E. Nel. Polyethyleneimine coating enhances the cellular uptake of mesoporous silica nanoparticles and allows safe delivery of siRNA and DNA constructs. *ACS Nano*, 3: 3273–3286, 2009.

[20] Chen, A.M., M. Zhang, D. Wei, D. Stueber, O. Taratula, T. Minko, and H. He. Co-delivery of doxorubicin and Bcl-2 siRNA by mesoporous silica nanoparticles enhances the efficacy of chemotherapy in multidrug-resistant cancer cells. *Small*, 5: 2673–2677, 2009.

[21] Meng, H., M. Liong, T. Xia, Z. Li, Z. Ji, J.I. Zink, and A.E. Nel. Engineered design of mesoporous silica nanoparticles to deliver doxorubicin and P-glycoprotein siRNA to overcome drug resistance in a cancer cell line. *ACS Nano*, 4: 4539–4550, 2010.

[22] Hom, C., J. Lu, M. Liong, H. Luo, Z. Li, J.I. Zink, and F. Tamanoi. Mesoporous silica nanoparticles facilitate delivery of siRNA to shutdown signaling pathways in mammalian cells. *Small*, 6: 1185–1190, 2010.

[23] Bhattarai, S., E. Muthuswamy, A. Wani, M. Brichacek, A. Castañeda, S. Brock, and D. Oupicky. Enhanced gene and siRNA delivery by polycation-modified mesoporous silica nanoparticles loaded with chloroquine. *Pharmaceutical Research*, 27: 2556–2568, 2010.

[24] Kam, N.W.S., Z. Liu, and H. Dai. Functionalization of carbon nanotubes via cleavable disulfide bonds for efficient intracellular delivery of siRNA and potent gene silencing. *Journal of the American Chemical Society*, 127: 12492–12493, 2005.

[25] Hirsch, A. Functionalization of single-walled carbon nanotubes. *Angewandte Chemie International Edition in English*, 41: 1853–1859, 2002.

[26] Elbashir, S.M., J. Harborth, W. Lendeckel, A. Yalcin, K. Weber, and T. Tuschl. Duplexes of 21-nucleotide RNAs mediate RNA interference in cultured mammalian cells. *Nature*, 411: 494–498, 2001.

[27] Yang, R., X. Yang, Z. Zhang, Y. Zhang, S. Wang, Z. Cai, Y. Jia, Y. Ma, C. Zheng, Y. Lu, R. Roden, and Y. Chen. Single-walled carbon nanotubes-mediated *in vivo* and *in vitro* delivery of siRNA into antigen-presenting cells. *Gene Therapy*, 13: 1714–1723, 2006.

[28] Zhang, Z., X. Yang, Y. Zhang, B. Zeng, S. Wang, T. Zhu, R.B.S. Roden, Y. Chen, and R. Yang. Delivery of telomerase reverse transcriptase small interfering RNA in complex with positively charged single-walled carbon nanotubes suppresses tumor growth. *Clinical Cancer Research*, 12: 4933–4939, 2006.

[29] Wang, X., J. Ren, and X. Qu. Targeted RNA interference of cyclin A2 mediated by functionalized single-walled carbon nanotubes induces proliferation arrest and apoptosis in chronic myelogenous leukemia K562 cells. *ChemMedChem*, 3: 940–945, 2008.

[30] Krajcik, R., A. Jung, A. Hirsch, W. Neuhuber, and O. Zolk. Functionalization of carbon nanotubes enables noncovalent binding and intracellular delivery of small interfering RNA for efficient knock-down of genes. *Biochemical and Biophysical Research Communications*, 369: 595–602, 2008.

[31] Podesta, J.E., K.T. Al-Jamal, M.A. Herrero, B. Tian, H. Ali-Boucetta, V. Hegde, A. Bianco, M. Prato, and K. Kostarelos. Antitumor activity and prolonged survival by carbon-nanotube-mediated therapeutic siRNA silencing in a human lung xenograft model. *Small*, 5: 1176–1185, 2009.

[32] Bartholomeusz, G., P. Cherukuri, J. Kingston, L. Cognet, R. Lemos, T.K. Leeuw, L. Gumbiner-Russo, R.B. Weisman, and G. Powis. In vivo therapeutic silencing of hypoxia-inducible factor 1 alpha (HIF-1alpha) using single-walled carbon nanotubes noncovalently coated with siRNA. *Nano Research*, 2: 279–291, 2009.

[33] Ladeira, M.S., and *et al*. Highly efficient siRNA delivery system into human and murine cells using single-wall carbon nanotubes. *Nanotechnology*, 21: 385101, 2010.

[34] Al-Jamal, K.T., L. Gherardini, G. Bardi, A. Nunes, C. Guo, C. Bussy, M.A. Herrero, A. Bianco, M. Prato, K. Kostarelos, and T. Pizzorusso. Functional motor recovery from brain ischemic insult by carbon nanotube-mediated siRNA silencing. *Proceedings of the National Academy of Sciences*, 108: 10952–10957, 2011.

[35] Foillard, S., G. Zuber and E. Doris. Polyethylenimine-carbon nanotube nanohybrids for siRNA-mediated gene silencing at cellular level. *Nanoscale*, 3: 1461–1464, 2011.

[36] Libutti, S.K., G.F. Paciotti, A.A. Byrnes, H.R. Alexander, W.E. Gannon, M. Walker, G.D. Seidel, N. Yuldasheva, and L. Tamarkin. Phase I and pharmacokinetic studies of CYT-6091, a novel PEGylated colloidal gold-rhTNF nanomedicine. *Clinical Cancer Research*, 16: 6139–6149, 2010.

Index

About the Authors

Dr. Surendra Nimesh is an internationally recognized expert of nanotechnology for biological applications with specialization in drug and gene delivery. He received his M.S. in Biomedical Science from Dr. B.R. Ambedkar Center for Biomedical Science Research (ACBR), University of Delhi, Delhi, India in 2001. He completed his PhD. in Nanotechnology with Prof. Ramesh Chandra at ACBR and Dr. K.C. Gupta at Institute of Genomics and Integrative Biology (CSIR), Delhi, India in 2007. After completing his postdoctoral studies with Prof. M.D. Bushmann at Ecole Polyetchnique of Montreal, Montreal in 2009, he joined Clinical Research Institute of Montreal, Montreal, Canada as Postdoctoral Fellow. Presently, he is working with Prof. Satya Prakash at McGill University, Montreal, Canada. He has authored 14 research papers, 2 review articles in international peer reviewed journal and 4 book chapters. He has given two invited lectures in international conferences in Turkey in 2010 and presented several posters in international conferences in India and US. His biography has been listed in Marquis Who's Who in Science and Engineering 2011–2012 (11th Edition). He has received Council of Scientific and Industrial Research- University Grants Commission's, India Junior Research Fellowship (JRF) and Eligibility for Lectureship — National Eligibility Test (NET), Dec. 2000 and Young Scientist Project from Department of Science and Technology (SERC Fast Track Proposals for Young Scientists Scheme), Government of India, March 2007. Apart from this, during his doctoral degree and research work, he had experience in training masters and undergraduate interns and made several presentations in the institute and abroad. His research interests include nanoparticles mediated gene, siRNA and drug delivery for therapeutics.

 Prof. Ramesh Chandra is a distinguished scientist and an outstanding researcher in the field of Biomedical Sciences. He is the Founder Director of the Dr. B. R. Ambedkar Center for Biomedical Research at the University of Delhi and has been the Vice-Chancellor of the Bundelkhand University, Jhansi (1999–2005) as well as the President of the Indian Chemical Society (2004–06) and Member, Planning Commission, Government of U.P., India.

Prof. Chandra shows deep commitment to the cause of higher education and research and possess in ample measure, quality of dynamic leadership and a vision required to build academic institutions. Prof. Chandra started his research career at the University of Delhi. He went to the New York Hospital-Cornell University Medical Center and the Rockefeller University, New York and State University of New York at Stonybrook. He conducted advanced research at the Harvard University Medical School-Massachusetts General Hospital, jointly at MIT, Cambridge USA. Over the last 33 years, Prof. Chandra has contributed largely in the field of Chemical Sciences and particularly in New Drug Discovery and Development as well as Drug Metabolism. His research work is being used in the development of drugs for physiological jaundice/ Neonatal Jaundice and development of chemotherapeutic agents for the treatment of breast and ovarian cancers, and drugs for diabetes and hypertension. He has supervised 81 Ph. D., 10 M. Phil. and in all trained more than 100 research scholars, published nearly 225 original Scientific Research Papers including Review Articles/ Monographs in International journals of repute.

Prof. Chandra is the recipient of several professional national/ international recognitions. These includes Vidya Ratan Gold Medal (2005); Dr. BR Ambedkar National Award (2004); Bronze Medal of the Chemical Research Society of India (CRSI) (2004); Lifetime Achievement Award of The Indian Chemical Society (ICS) (2003); Prof. Ghanshyam Srivastava Commemoration Award of ICS (2002); Rajib Goyal Award for Young Scientists (2002);

Prof. D. P. Chakraborty Commemoration Award of ICS (2001); Award of the Highest Honor of Soka University, Tokyo, Japan (2000); IMNM-99 Award and Gold Medal in Integrated Medicine for New Millennium (1999); J William Fulbright Scholarship(1993); UGC Research Scientist Award (1988); UGC Career Award(1993); The Rockefeller Foundation USA-Biotechnology Career Award (1993). He is Fellow of The Royal Society of Chemistry, London; International Academy of Physical Sciences; Institution of Chemists, India and the Indian Chemical Society. He is member of several International Scientific Societies.

He has been a member of the Governing Council, BOG, Executive/ Academic Councils of several Universities/ Institutions and Member, U.P. State, Youth Welfare Council and Member, Council of Higher Education, U.P. He is also a Consultant and Advisor to the various multinational companies like Polaroid Corporation, Diakron Pharmaceuticals, USA, HIKMA Pharmaceuticals-Jordan and Director of BIZ SHAKTI, India etc. and also Director of PSU, Govt. of India. Prof. Chandra is a prolific writer and displays extraordinary flair for writing on themes particularly to Higher Education.

Technology in Biology and Medicine

Other books in this series:

Volume 1
Adult Stem Cell Standardization
Editor: Paolo Di Nardo
ISBN: 978-87-92329-74-5

For Product Safety Concerns and Information please contact our EU
representative GPSR@taylorandfrancis.com
Taylor & Francis Verlag GmbH, Kaufingerstraße 24, 80331 München, Germany

www.ingramcontent.com/pod-product-compliance
Ingram Content Group UK Ltd.
Pitfield, Milton Keynes, MK11 3LW, UK
UKHW052300180425
457613UK00009B/267